Ordinal Optimization:
Soft Optimization for Hard Problems

Wisdom consists of knowing when to avoid perfection.
— Horowitz's Rule[*]

---

[*] *The Complete Murphy's Law* by Arthur Bloch, Price Stern Sloan Publisher, Los Angeles, 1991.

ORDINAL OPTIMIZATION

SOFT OPTIMIZATION FOR HARD PROBLEMS

Yu-Chi Ho
Harvard University
Massachusetts, USA
Tsinghua University
Beijing, China

Qian-Chuan Zhao
Tsinghua University
Beijing, China

Qing-Shan Jia
Tsinghua University
Beijing, China

Yu-Chi Ho, PhD, Professor
Harvard University
Cambridge, MA, USA
Tsinghua University
Beijing, China
e-mail: ho@deas.harvard.edu

Qian-Chuan Zhao, Ph.D.
Professor of Automation Engineering
Center for Intelligent & Networked
    Systems
Tsinghua University
Beijing, China
e-mail: zhaoqc@tsinghua.edu.cn

Qing-Shan Jia, PhD, Lecturer
Center for Intelligent & Networked Systems
Tsinghua University
Beijing, China
e-mail: jiaqs@tsinghua.edu.cn

Ordinal Optimization: Soft Optimization for Hard Problems
by Yu-Chi Ho, Qian-Chuan Zhao and Qing-Shan Jia

ISBN-13: 978-1-4419-4243-2   e-ISBN-13: 978-0-387-68692-9

Printed on acid-free paper.

9 8 7 6 5 4 3 2 1

springer.com

To Sophia,
You have made life worth living and ideas possible.

———Larry

To Betty,
You have made life simple and beautiful.

———Qian-Chuan

To Huai-Jin Jia and Guo-Hua Zhou,
You have made my life possible.

———Qing-Shan

Table of Contents

Appendix A Fundamentals of Simulation
and Performance Evaluation

Appendix B Introduction to Stochastic Processes
and Generalized Semi-Markov Processes as Models
for Discrete Event Dynamic Systems
and Simulations

Appendix C Universal Alignment Tables
for the Selection Rules in Chapter III

Preface

This book is a research monograph dealing with the subject of perfor-mance evaluation and optimization for complex systems. At the same time, it can be used and has been used as the main textbook for a first-year graduate course at Harvard University and Tsinghua University for the past 15 years. Exercises are included throughout the book.

According to the authors' experience, engineering methodology or tools can be more easily taught and accepted when it is applied in a specific area no matter how narrow or broad it may be. In this way, students usually have more incentives to study further and are more likely to better appreciate the importance of and the raison d'etre for various features of the methodology. Such an appreciation of "what is essential" is a useful skill for students to acquire regardless of their career choice later on.

This book focuses on the performance evaluation and optimization of complex human made systems, which are known as Discrete Event dynamic Systems (DEDS), such as automated manufacturing plants, large communi-cation networks, traffic systems of all kinds, and various paper processing bureaucracies, etc. There is no simple analytical model for the description of such systems, such as differential equations, since such systems are mostly developed according to human made rules of operation rather than physical laws. Brute force simulation models are often the only choice. If traditional methods are used, the computational burden relating to evaluation and optimization of simulation models often renders their solutions computa-tionally infeasible. Ordinal Optimization (OO) is a methodology invented in early 1990s to get around this fundamental difficulty. After more than 14 years' development, it becomes a complete methodology covering all the aspects of applying the methodology to practical problems. A large number of works on the subject are ready for reference (Shen et al. 2007). A book collecting all the information in one place seems to be in order.

Only basic knowledge of probability and an undergraduate background in engineering are needed for reading this book. This book is divided into eight Chapters. The first chapter introduces the rationale and the last one provides application examples. In chapter I, it discusses the computational difficulty in performance evaluation and optimization via simulation models when using traditional means, followed by an introduction of the basic ideas and concepts regarding Ordinal Optimization (OO) in Chapter II. From Chapters III to VII, we provide various extensions of the OO methodology, by which the practical problems may be solved more or less routinely. For readers who are not familiar with simulation and modeling of DEDS, the book provides the fundamentals of simulation and stochastic processes in Appendices A and B. Appendix C contains data and parameters needed in Ch. III. Additional exercises are provided in Appendix D.

The structure of this book is elaborated in Fig. 1.

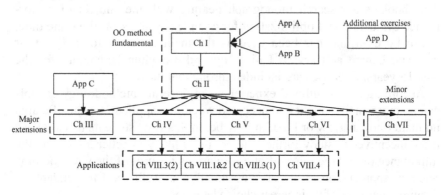

Fig. 1. Structure of the book

Acknowledgements

This book owes its existence to the National Science Foundation, the Air Force Office of Scientific Research, the Office of Naval Research and the Army Research Office of the United States. Without their more than 40 years of visionary and unfettered support, especially to my researches on ordinal optimization in the last 15 years, the publication of this book would not be possible. Finally, I take this opportunity also to thank all my Ph.D. students (53 in total) for their ideas and materials in support of my research. I've also learned a lot from them.

Yu-Chi Ho, March 2007, Cambridge, MA USA

I would like to thank Fundamental Research Funds from Tsinghua University, the Chinese Scholarship Council, National Natural Science Foundation of China (60274011 and 60574067), Chinese NCET program (NCET-04-0094), the National 111 International Collaboration Project and United Technology Research Center. I would also like to express my gratitude to Prof. Da-Zhong Zheng and Prof. Yu-Chi Ho. for their insightful suggestions to improve the work.

Qian-Chuan Zhao, March 2007, Beijing, China

This book could not have been done without the help of a large number of organizations, professors and students. In particular, I want to thank National Natural Science Foundation of China (60274011 and 60574067), Chinese NCET program (NCET-04-0094) and National Basic Research Program of China (973 project) (contract 2004CB217900) and United Technology Research Center for for supporting me to do the work since 2001. I would especially like to mention Prof. Ho, Prof. Zhao, Prof. Xiaohong Guan, Hongxing Bai, Zhen Shen, and Yanjia Zhao for their advice and comments from which I have benefited a lot. A word of thanks to my parents for their love and encouragement.

Qing-Shan Jia, March 2007, Beijing, China

Chapter I Introduction

"Optimization", taken in the broadest sense as seeking for improvement, is an idea as old as mankind itself. In fact, it can be argued that it is the "**raison d'être**" for our civilization. Without the desire to improve, progresses on all fronts will stall. Yet the study of optimization as a discipline but not as individual endeavors on specific problems did not begin until the invention of calculus, which enabled the mathematical modeling of a large number of physical phenomena. The theory of maxima/minima and convexity emerged as a result. Yet the numbers of real world problems that can be explicitly solved by mathematics alone remain limited until the development of the computer. Suddenly, many algorithms, which were previously thought to be infeasible for the numerical and iterative solution of difficult optimization problems, now become possible. The golden age of optimization took off in the 1950s and is still ongoing.

On the other hand, in spite of the tremendous development of the science and art of optimization and computation, there remain many problems that are still beyond our reach. Among them are the class of combinatorial NP-hard problems and the well known "curse of dimensionality" in dynamic programming. Exponential growth is one of the problems that mathematics and computers cannot overcome. Furthermore, the computational burden of a problem does not always necessarily arise because of problem size. Complexity of a problem can also impose infeasible computational burdens. We have here in mind, the class of computational problems that arise out of **simulation** models.

Civilization have increasingly created complex human-made systems, such as large-scale electric power grids, air and land traffic control systems, manufacturing plants and supply chains, the Internet and other communication networks, etc. Such systems operate and evolve in time via human- made rules of operation, and are difficult to describe and capture by succinct mathematical models such as differential equations for physical systems. Collectively, they are known as Discrete Event Dynamic Systems (DEDS) (Ho 1989). In fact, the only faithful description of such complex systems is by way of an electronic copy, i.e., a simulation model/program that duplicates everything the real system does in real or simulation time. Evaluation of the performance of such systems is accomplished by running such simulation models rather than experimenting with the real systems. This is important for design purposes when the real system does not even exist or when it is inconvenient or impossible to do experiments on the real system. But having a simulation model is not the end of the problem difficulty. Experiments with such model are often quite time consuming. In Table 1.1 below, we listed several typical real problems (some will be discussed in detail in Chapter VIII) as well as the typical time required for one particular evaluation of their performance metric. We are not attaching any particular significance to Table 1.1 except to convey a sense of the time required for performance evaluation and optimization using simulation models and computers available today. The time shown in each row is either for a single performance evaluation or one run (replication) of the simulation model.

Table 1.1. The time to simulate some real systems

Real system	Performance	Simulation time
Remanufacturing system	Accurately evaluate the average cost of a maintenance strategy by 1000 independent replications	30 minutes
Congestion control and buffer management in a computer network	A NS[1]-2-based simulation of the 1000-second dynamics of a 12000-node-single-bottleneck computer network	1.5 hours
Security evaluation and optimization for a large scale electric power grid	A simulation of the 30-second dynamics of a large scale electric power grid with 5000 buses and 300 generators after a failure event	2 hours
Scheduling of a transportation network	A simulation of the 24-hour dynamics of a transportation network with 20 intersections	2 hours
Turbine blade design problem	A 3D extrusion simulation with Finite Element Methodology	7 days

From Table 1.1, we can get a feel of the immense computational burden particularly if optimization in addition is attempted with such models via successive iteration. Mathematically, we can represent the performance of a simulation model/program as $L(x(t;\theta,\xi))$ where $x(t;\theta,\xi)$ is the trajectory (sample path) of the system as the simulation evolves in time; $\theta \in \Theta$, the various system parameters that may be subject to design choices; Θ is the

---

[1] NS is short for Network Simulator, which is a software to simulate the computer network.

search or design space for θ; ξ, all the randomness that may occur in the system as it evolves during a particular sample trajectory; and L, a functional of $x(t;\theta,\xi)$ that defines the performance metric of the system we are interested in.

Because of the inevitable randomness that occurs in such real systems, we work with expected performance by defining

$$J(\theta) \equiv E\left[L\left(x(t;\theta,\xi)\right)\right], \qquad (1.1)$$

where the expectation is taken with respect to the distribution of all the randomness, ξ. Computationally, Eq. (1.1) is calculated by

$$J(\theta) \equiv E\left[L\left(x(t;\theta,\xi)\right)\right] \approx \frac{1}{n}\sum_{t=1}^{n}L\left(x(t;\theta,\xi_i)\right) \qquad (1.2)$$

where n is usually a large number in practice and ξ_i denotes the sample randomness during the i-th replication of the simulation. It is well known that the accuracy of the estimate of J using a finite n improves as $1/(n)^{1/2}$ ((Fishman 1996; Yakowitz 1977; Gentle 1998; Landau and Binder 2000) as well as see Chapter II). Here comes the computational difficulty. For every *one* order of magnitude increases in accuracy of the estimate for J, n must increase by *two* orders of magnitude. Since each sample run of the simulation model of a complex system may consume considerable time, running the simulation model n times to achieve an accurate estimate of J may impose a heavy burden[2]. As a result, simulation models are often used for validation of a design obtained by other means but not for optimization purposes.

To make matters worse, while the literature on optimization and decision-making is huge, much of the concrete analytical results are associated with what may be called **Real Variable Based methods**. The idea of successive approximation to an optimum (say, minimum) by sequential improvements based on local information is often captured by the metaphor of "skiing downhill in a fog". The concepts of gradient (slope), curvature (valley), and trajectories of steepest descent (fall line) all require the notion of derivatives and are based on the existence of a more or less smooth response surface. There exist various first and second order algorithms of feasible directions for the iterative determination of the optimum (minimum) of an arbitrary multi-dimensional response or performance

---

[2] As the examples show in Table 1.1.

surface[3]. Considerable numbers of major success stories exist in this genre including the Nobel Prize winning work on linear programming. It is not necessary to repeat or even reference these here.

On the other hand, we submit that the reason many real world optimization problems remain unsolved is partly due to the changing nature of the problem domain, which makes calculus or real variable based method less applicable. For example, a large number of human-made system problems mentioned above involve combinatorics, symbolic or categorical variables rather than real analysis, discrete instead of continuous choices, and synthesizing a configuration rather than proportioning the design parameters. Optimization for such problem seems to call for general search of the performance terrain or response surface as opposed to the "skiing downhill in a fog" metaphor of real variable based performance optimization[4]. Arguments for change can also be made on the technological front. Sequential algorithms were often dictated as a result of the limited memory and centralized control of earlier generations of computers. With the advent of modern massively parallel machines, distributed and parallel procedures can work hand-in-glove with **Search Based method** of performance evaluation (Thomke 2003).

The purpose of this book is to address the difficulties of optimization problems described above – the optimization of complex systems via simulation models or other computation-intensive models involving possible stochastic effects and discrete choices. As such, the book is complementary to existing optimization literature. The tools to be described here do not replace but can be used separately or in conjunction with other methodological tools of optimization.

If we accept the need for search based method as complement to the more established analytical real variable techniques, then we can next argue that to quickly narrow the search for optimum performance to a "good enough" subset in the design universe is more important than to estimate accurately the values of the system performance during the process of optimization. We should compare "order" first and estimate "value" second, i.e., **ordinal optimization** comes before **cardinal optimization**. Furthermore, we shall argue that our preoccupation with the "best" may be only an ideal that will always be unattainable or not cost effective. Real

---

[3] We include here also the more recently developed gradient tool of **Perturbation Analysis** for discrete event systems.

[4] We hasten to add that we fully realize the distinction we make here is not absolutely black and white. A continuum of problems types exists. Similarly, there is a spectrum of the nature of optimization variables or search space ranging from continuous to integer to discrete to combinatorial to symbolic.

world solution to real world problem will involve compromise for the "good enough"[5]. The purpose of this book is to establish the distinct advantages of this softer and ordinal approach for the search based type of problems, to analyze some of its general properties, and to show the several orders of magnitude improvement in computational efficiency that is possible under this mind set.

Of course, this softer optimization is not unique with our ordinal approach to be explained below. Heuristics of various kind (Glover 1989) used in successfully solving NP-hard combinatorial problems and the huge literature on fuzzy control (Passino and Yurkovich 1998; Driankov et al. 2006) all advocate concepts that can be grouped under the rubric of soft optimization. It is not the primary intention of this book to offer an alternative to these established subject matters. On the other hand, ordinal optimization as presented in this book does offer a *quantified* rather than just heuristic account of reducing the intensive computational burden associated with the performance evaluation and optimization of complex human made systems.

---

[5] Of course, here one is mindful of the business dictum, used by Mrs. Fields for her enormously successful cookie shops, which says, "Good enough never is". However, this can be reconciled via the frame of reference of the optimization problem.

Chapter II Ordinal Optimization Fundamentals

1 Two basic ideas of Ordinal Optimization (OO)

There are two basic ideas in OO:

- "Order" is much more robust against noise than "Value"
- Don't insist on getting the "Best" but be willing to settle for the "Good Enough"

Of course readers may rightly point out that these ideas are hardly new. Good engineers and designers do this all the time when confronted with difficult and complex problems of performance evaluation and optimization. Our contribution in this book is simply that we have developed a theory to **quantify** these two ideas. The practice of these two ideas is now knowledge-based instead of being experience-based. The expertise of a good designer acquired from experience will now be available to everyone who uses the tools discussed in this book. Moreover, the user will have numerical measures rather than just gut feelings. The quantification will come later in the book. For now let us simply explain the ideas in intuitive terms.

Idea No.1 Order is easier than Value. Imagine you hold two identically looking boxes with unknown content in your two hands. You are asked to judge which one is heavier than the other – an "ordinal" question. Almost all of us can correctly tell the answer, even if there is only a very slight difference between the weights of the two boxes. But if we are asked to estimate the difference in weight between the two boxes – a "cardinal" question, we will have a hard time. This is the essence of the first tenet of OO. In fact later on in this chapter we shall prove that "Order" converges exponentially fast in stochastic simulation models against the $1/(n)^{1/2}$ convergence rate of "Value" or "Confidence Interval" as discussed in chapter I.

An intuitive and graphical illustration of this idea is shown in Fig. 2.1.

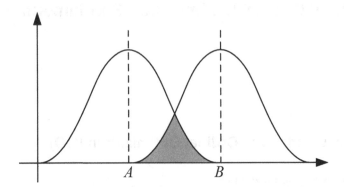

Fig. 2.1. Comparison of two values *A* and *B* corrupted by noise

We have two values *A* and *B*. But we can only observe the values through noises, which are zero mean and normally distributed. If we compare samples of these noisy values of *A* and *B*, then we will reach an incorrect answer, i.e., $A > B$, only when the samples of *A* is larger than that of *B*. This happens basically when the samples fall in the shaded triangle area. This area decreases rapidly as the values of *A* and *B* are drawn apart.

Idea No.2 Nothing but the best is very costly. The second tenet of OO rests on the idea of relaxing the goal of performance evaluation. We are asked to retreat from "nothing but the best" to a softer goal of being "good enough", e.g., settle for anything in the top-*g* choices. The small retreat can buy us quite a bit in the easing of the computational burden. Again the quantification will come later on in the chapter. However, an intuitive explanation is helpful to fix ideas at this point. When we are searching only for the best in Θ, there is only one solution (unless of course in the less common cases when there are multiple optima), and we can easily fail (i.e., the truly best design is not the observed best one). But if we are willing to settle for any one of the top-2 choices, we fail only when neither design is within the observed top-2, and if we are willing to settle for any of the top-*g* choices, there will be at least *g*! satisfactory alternatives, the possibilities increases superlinearly. More technically, for $g = 2$, the probability of which is almost the square of the previous failure probability for a large enough Θ; for top-3, we fail only when none of these three designs is within the observed top-3, the probability of which is almost the cube of the first failure probability; and for top-*g*, where *g* is small compared to the size of Θ, we fail only when none of these *g* designs is within the observed top-*g*, the probability of which is almost the first failure probability to the power of *g*.

Exercise 2.1: Prove the above statements concerning the failure probabilities.

This exponential decrease of failure probability (thus the exponential increase in successful probability) contributes to significant decrease in search, and hence computational cost. Another way of clarifying the advantage of this idea is to observe that independent noisy observation or estimation error may worsen as well as help the performance "order" of a particular choice of θ. Thus, as a group, the top-n choices can be very robust against perturbations in order so long as we don't care about the exact order within this group.

Lastly, it is not hard to convince oneself that the ease of computational burden through these two ideas is not additive but multiplicative since they are separate and independent factors. Together, as we shall demonstrate throughout this book, they produce orders of magnitude improvement in efficiency. Problems previously thought to be infeasible computationally are now within reach.

2 Definitions, terminologies, and concepts for OO

We first introduce some definitions, which will be used throughout this book

θ The various system parameters that may be subject to design choices.

Θ The search space for the optimization variables θ. We can simply assume it to be a very large but finite set consisting of zillions of choices.

J The performance criterion for the system as already defined in Eqs. (1.1) and (1.2).

$\hat{J}$ The estimated or observed performance for the system. This is usually done by using a crude but computationally easy model of the system performance.

w The observation noise/error, which describes the difference between the true performance and the estimated/observed performance, i.e., $\hat{J}(\theta) = J(\theta) + w$.

ξ All the random variables with known distribution used in the simulation model.

$x(t;\boldsymbol{\theta},\boldsymbol{\xi})$ A sample trajectory of the system given the design θ and the realizetion ξ. In simulation terminology, this is often called a **replication**.

L A functional of $x(t;\, \theta,\xi)$ that defines the performance metric of the system on a sample path. The system's performance criterion is given by the expectation $J = E[L(x(t;\, \theta,\xi))]$.

N The number of designs uniformly chosen in Θ. It is understood that for each design θ, there corresponds a value of J as defined in Eqs. (1.1) and (1.2). When there is no danger of confusion we also use N to denote the set of designs chosen.

G The Good Enough set, usually the top-g designs of Θ or of N.

S The set of selected designs in N, usually the estimated top-s of N.

Selection Rule

The method that is used to determine the set S, e.g., it could be **blind pick** or **horse race**[1] or other rules using the estimated performance to order the designs (based on a crude model).

$G\cap S$ The truly good enough designs in S.

The relationship among Θ, G, S and the true and estimated optimal design are shown in Fig. 2.2. Note the sets G and S in OO play the analogous role as the true optimum and the estimated optimum respectively in regular optimization.

AP, Alignment Probability $\equiv$ Prob[$|G\cap S|\geq k$]

The probability that there are actually k truly good enough designs in S (represented by the dotted area in Fig. 2.2 above). k is called the **alignment level**. This number quantifies how a crude model can help to assure the determination of "good enough" designs.

OPC, Ordered Performance Curve

A (conceptual) plot of the values of J as a function of the order of performance, i.e., the best, the second best, and so on. If we are minimizing then the OPC must be a non-decreasing curve[2].

---

[1] The blind choice selection rule is to arbitrarily and randomly selected s θ's from N as the set S. The horse race selection method is simply to compare the observed performance values, $\hat{J}_1, \hat{J}_2, \ldots, \hat{J}_N$ and select the designs with top-s observed values as the set S. More details can be found in Chapter III.

[2] In this book, we usually deal with finite but a large number of designs. For visualization, we show continuous curves in Fig. 2.3 and in some of the following figures, instead of a large number of discrete points. These figures can be regarded as limits of discrete points with high density.

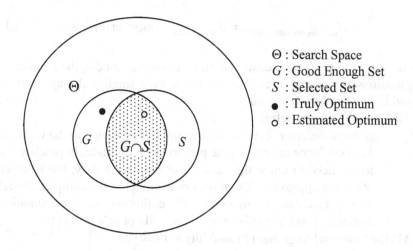

Fig. 2.2. Graphical Illustration of Θ, *G*, and *S*

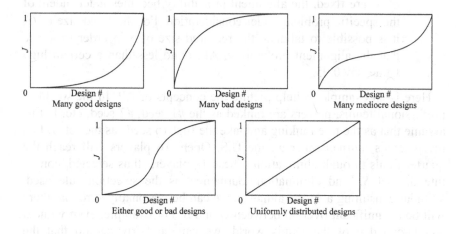

Fig. 2.3. Different types of OPC

c   denotes the class or type of OPC for the problem. For non-decreasing curves, there are only five general types as shown in Fig. 2.3 below.

The fact that "order" is robust against noise (as will be shown shortly) gives us the possibility that we may use a crude model to determine the order of various designs. In other words, we visualize and set up

$$J_{\text{complex simulation model}} = J_{\text{crude model}} + \text{noise/error}\,^3. \qquad (2.1)$$

In terms of Eq. (1.2) for the simulation of a complex model, the number of replications, n, is very large. However n is very small and may even be equal to one for the crude model.

Given Eq. (2.1), we define

σ^2 as the noise/error level. In simulation, this is related to the width of the confidence interval of the performance estimate. In practice we do not need to know the value of σ^2 very accurately, but whether or not the approximation of the crude model to the complex model is very bad, bad, or moderate. By definition the approximation cannot be good. For otherwise, there will not be a problem.

UAP, the Universal Alignment Probability $\equiv \text{Prob}\left[\,|G \cap S| \geq k/N, \sigma^2, C\,\right]$
$\equiv \textbf{UAP}(N, \sigma^2, C)$

We assert here, and will establish later (in Section 5), that once N, σ^2, C are fixed, the alignment probability becomes independent of the specific problem under consideration. For the fixed size of G, it is possible to tabulate the required size of S in order to insure that the alignment probability, AP, is no less than a certain high value, say 0.95.

Here is an example to help picture the concepts of AP. In the world of professional tennis, players are ranked as the #1 seed, #2 seed, etc. Let us assume that as the true ranking and take the first 16 seeds as the set G. In a given tennis tournament, e.g., the U.S. Open, 16 players will reach the quarter-finals through elimination. These 16 players, thus selected, constitute the set S. And elimination tournament is the selection rule used. Without consulting a sports almanac, we can be reasonably sure that there will be a significant overlap between G and S. For example, even without any knowledge of the tennis world, we can be fairly certain that the $\text{Prob}[|G \cap S| \geq 1] \approx 1$. Furthermore, such near certainty will exist in other sports such as horse race. Thus, we claim the use of the adjective "universal" for such APs which are tabulated as a function of N, σ^2, and C. In fact we shall show in later chapters how UAPs can be used to narrow down searches for the "good enough". There is also a simple demonstration

---

[3] It will be introduced later (in Section 6) that how to use OO to deal with different simulation models, i.e., both stochastic simulation models and deterministic complex simulation models. To cover both types of models, we use "noise/error" here. In Section 6 we will show that noise and error can be regarded as the same under the concept of Kolmogorov complexity.

explained immediately in Section 3 below which further illustrates the universality of the concept of alignment probability.

Finally, we summarize the spirit of ordinal optimization as:

> ***Instead of the best for sure, we seek***
> ***the good enough with high probability.***

3 A simple demonstration of OO

We present here a simple generic demonstration that everyone can do to convince themselves of the validity of the two basic tenets of OO as discussed in Section 1 and the universality of the alignment probability of OO. Let the search space Θ have 200 designs, say 1,2, ..., 200. Without loss of generality, we let choice #1 be the best, #2, the second best, and so on to #200; and the performances for simplicity be 1, 2, ..., 200[4]. Mathematically, we have $J(\theta_i) = i$ in which case the best design is θ_1 (we are minimizing) and the worst design is θ_{200} and the Ordered Performance Curve (OPC) is linearly increasing. For any finite i.i.d noise w, we can implement $\hat{J}(\theta) = J(\theta) + w$ and directly observe these performances through noisy measurements as in Eq. (2.1) with noise distributed as $U[0,100]$ or $U[0,10000]$. Let the good enough performance G be 1, . . ., 12 (the top-6%), and S be the observed top-6%. We are interested in $|G \cap S|$.

All of these can be simply implemented in a spreadsheet with Θ represented by 200 rows as in Fig. 2.4.

Design # = θ	True performance $J(\theta)$	Noise $w \in U[0,W]$	Observed performance $J(\theta)+w$
1	1.00	87.98	88.98
2	2.00	1.67	3.67
.	.	.	.
.	.	.	.
.	.	.	.
199	199.00	32.92	231.92
200	200.00	24.96	224.96

Sort on this column
in ascending order

Fig. 2.4. Spread sheet implementation of generic experiment

---

[4] Actually any monotonically increasing numerical sequence will do.

Column 1 models the N (=200) alternatives and the true order 1 through N. Column 2 shows the linearly increasing OPC from 1 through N (=200). The rate of OPC increase with respect to the noise variance σ^2 essentially determines the estimation or approximation error of $\hat{J}(\theta)$. This is shown by the random noise generated in column 3 which, in this case, has a large range $U[0,100]$ or $U[0,10000]$. Column 4 displays the corrupted (or estimated or observed) performance. When we sort on column 4, we can directly observe the alignment in column 1, i.e., how many numbers 1 through g (=12) are in the top-g rows. For example, in Fig. 2.5 we show what the sorting result of the table in Fig. 2.4 may look like. Try this and you will be surprised! It takes less than two minutes to setup on Excel.

Design # = θ	True performance $J(\theta)$	Noise $w \in U[0,W]$	Observed performance $J(\theta)+w$
2	2.00	1.67	3.67
5	5.00	20.12	25.12
.	.	.	.
.	.	.	.
.	.	.	.
90	90.00	79.09	169.09
193	193.00	90.85	283.85

Fig. 2.5. The spread sheet in Fig. 2.4 after sorting on the observed performance in ascending order (Note: Column 4 is completely sorted in ascending order)

An already implemented version can also be found on http://www.hrl. harvard.edu/~ho. One can also simulate the result of blind pick by making the noise $U[0, 10000]$ and repeat the Excel simulation. It should be noted that the above demonstration is rather general. Except for the assumption of independent noise/error in Eq. (2.1), the results are applicable to any complex computational problems when approximated by a crude model. Chapters below will further discuss and exploit this generality.

At this point, except for establishing more carefully the validity of the two basic ideas of ordinal optimization (in the rest of this chapter) and other extensions (in the future chapters), we already have the procedure for the practical application of OO to complex optimization problems. This procedure is basically an elaboration of the demo above which we re-stated here for emphasis (Box 2.1):

Box 2.1. The application procedure of OO

Step 1: Uniformly and randomly sample N designs from Θ.
Step 2: Use a crude and computationally fast model to estimate the performance of these N designs.
Step 3: Estimate the OPC class of the problem and the noise level of the crude model. The user specifies the size of good enough set, g, and the required alignment level, k.
Step 4: Use Table 2.1 (in Section 5 below) to calculate $s = Z(g,k/\text{OPC}$ class, noise level).
Step 5: Select the observed top s designs of the N as estimated by the crude model as the selected set S.
Step 6: The theory of OO ensures that S contains at least k truly good enough designs with probability no less than 0.95.

Section 7 gives two examples of application of this procedure to complex problems. Interested reader may go directly to that section. Case studies of more complex real world problems can be found in chapter VIII.

4 The exponential convergence of order and goal softening

In this section, we provide the theoretical foundation of ordinal optimization method, namely, the alignment probability converges exponentially with respect to the number of replications (Section 4.2) and with respect to the sizes of good enough set and selected set (Section 4.3). The proofs can be viewed as a more formal explanation on the two tents of OO: optimization can be made much easier by order comparison and goal softening. As a preparation for the proof of exponential convergence w.r.t. order, we introduce the large deviation theory first (Section 4.1). The large deviation theory justifies why order comparison of two values A and B corrupted by noise is easy as we pointed out in Section 2.1. Readers who are not that interested in mathematical details and are willing to accept the intuitive idea introduced in Section 1 can skip or skim this section on a first reading.

The formal verifications in this chapter are given for the most frequently used selection rule—horse race rule. In OO, recalling that for the horse race rule, we are interested in the alignment probability (AP) that the observed top-s designs (estimated good enough designs) contain at least k of the actual top-g designs (real good enough designs).

To establish the exponential convergence properties formally, we need to introduce the problem formulation first. Assume that the N designs are indexed such that

$$J(\theta_1) < J(\theta_2) < J(\theta_3) < ... < J(\theta_N).$$

Let $L(\theta_i, n)$ be the sampled performance for the n-th replication and assume that for each design θ_i, $L(\theta_i, 1)$, $L(\theta_i, 2)$,..., $L(\theta_i, n)$, ... form a sequence of i.i.d. random variables with distribution such that for any $n>0$, $E[L(\theta_i, n)] = J(\theta_i)$. Let $\bar{\hat{J}}(\theta_i, n)$, $i=1,2,...,N$, be the performance estimates such that

$$\bar{\hat{J}}(\theta_i, n) = \frac{1}{n} \sum_{j=1}^{n} L(\theta_i, j) = J(\theta_i) + w(\theta_i, n),$$

where $w(\theta_i, 1)$, $w(\theta_i, 2)$, ..., $w(\theta_i, n)$, ... are estimation errors and $E[w(\theta_i, n)] = 0$.

4.1 Large deviation theory

Let us consider an i.i.d. sequence $x_1, x_2, ...$ with distribution function F (or density function f) and finite mean μ. In our context, the numbers in this sequence are observations of performance of a given design. Let $a > \mu$ and $b < \mu$ be two constants. The law of large numbers implies that

$$\text{Prob}\left[\frac{x_1 + \cdots + x_n}{n} \geq a\right] = \text{Prob}[x_1 + \cdots + x_n \geq na] \rightarrow 0 \text{ as } n \rightarrow \infty$$

and

$$\text{Prob}\left[\frac{x_1 + \cdots + x_n}{n} \leq b\right] = \text{Prob}[x_1 + \cdots + x_n \leq nb] \rightarrow 0 \text{ as } n \rightarrow \infty.$$

A fundamental question is: how fast do these two probabilities decrease?

Although it seems that this question is about a single design, it is important to us since there is a natural way to reduce our order comparison problem to it. In Fig. 2.6, the deviation probability $\text{Prob}\left[\dfrac{x_1 + \cdots + x_n}{n} \geq a\right]$ is

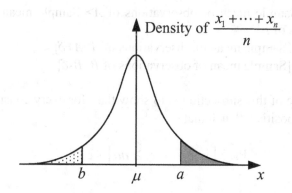

Fig. 2.6. Illustration of deviation probabilities

corresponding to the gray area and the deviation probability $\mathrm{Prob}\left[\dfrac{x_1 + \cdots + x_n}{n} \le b\right]$ the dotted area.

The connection between the deviation probabilities and the comparison of two fixed values A and B corrupted by noises can be interpreted as in Fig. 2.7 below. Assume $B > A$. Denote u as the position where the density functions of the two sample means meet. Then the rough estimation on misalignment probability (shaded area) in Fig. 2.1 (in Section 2.1) can be viewed as the sum of the gray area and the dotted area, where the gray area equals the deviation probability of A beyond u and the dotted area equals the deviation probability of B under u. Note u might change for a different n. A precise way of upper bounding the misalignment probability is to fix an amount of deviation less than or equal to $\delta = \dfrac{B-A}{2}$:

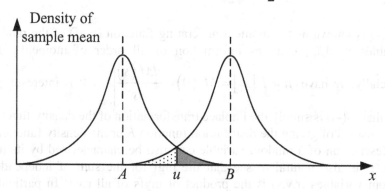

Fig. 2.7. Comparison of two values A and B corrupted by noises

Prob[Sample mean of observations of *A*> Sample mean of observations of *B*]

≤ Prob[Sample mean of observations of *A*>*A*+δ]

+ Prob[Sample mean of observations of *B*<*B*-δ].

The purpose of this subsection is to show that, for every constant $a > \mu$, there exists a positive β such that

$$\text{Prob}\left[x_1 + \cdots + x_n \geq na\right] \leq e^{-n\beta} \tag{2.2}$$

and, for each constant $b<\mu$

$$\text{Prob}\left[x_1 + \cdots x_n \leq nb\right] \geq e^{-n\beta}. \tag{2.3}$$

This implies that the probability for the sample mean $\dfrac{x_1 + \cdots + x_n}{n}$ to have finite deviation ("large deviation") from its mean decays exponentially. In the following, we shall show Eq. (2.2) and leave the similar justification of Eq. (2.3) to the readers. It is useful to define

$$M(s) \equiv E\left[e^{sx_1}\right] = \int e^{sy} dF(y) = \int e^{sy} f(y) dy.$$

Exercise 2.2: Let $x_1, x_2, \ldots$ be i.i.d. standard normal random variables, then please derive

$$M(s) = \frac{1}{\sqrt{2\pi}} \int e^{sy} e^{-\frac{1}{2}y^2} dy = e^{\frac{1}{2}s^2}.$$

$M(s)$ is known as the moment generating function (mgf) of the random variables x_i. $M(s)$ contains information of all order of moments, and especially we have $\mu = E[x_1] = M'(0) = \dfrac{dM(s)}{ds}\bigg|_{s=0}$. It is interesting to note that $M(-s)$ is simply the Laplace transformation of the density function f. So, instead of giving the distribution function F or the density function f, the description of a random variable can also be characterized by its mgf $M(s)$. It is also natural to see that the mgf for the sum of independent random variables (r.v.s) is the product of mgfs of all r.v.s. In particular,

$$\int e^{sy} dF_{x_1+\cdots+x_n}(y) = \left(\int e^{sy} dF(y)\right)^n = [M(s)]^n, \quad \text{where} \quad F_{x_1+\cdots+x_n} \quad \text{is the}$$

distribution of the r.v. $x_1+\ldots+x_n$.

For $s \geq 0$, mgf has the advantage of providing upper bounds on probability of events. In fact, we have

$$\text{Prob}[x_1 \geq b] \leq e^{-sb} M(s). \tag{2.4}$$

To see why this is true, we make the following observation.

$$\int_{y \geq b} f(y) dy \leq \int_{y \geq b} e^{s(y-b)} f(y) dy \leq \int e^{s(y-b)} f(y) dy. \tag{2.5}$$

The first inequality in Eq. (2.5) follows from the fact that $e^{s(y-b)} \geq 1$ when $s \geq 0$ in the range of integration $y \geq b$. The second inequality of Eq. (2.5) is due to the fact that $e^{s(y-b)} f(y)$ is always non-negative (recall f is a density function) and integration of a positive function over the entire region $(-\infty,+\infty)$ is always no less than integration over a part $[b,+\infty)$ of it . Thus Eq. (2.4) follows from Eq. (2.5) by noting

$$\text{Prob}[x_1 \geq b] = \int_{y \geq b} f(y) dy$$

and

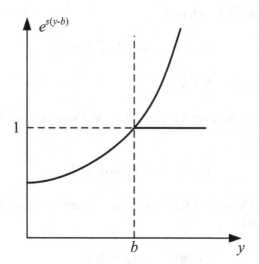

Fig. 2.8. Illustration of Eq. (2.5)

$$\int e^{s(y-b)} f(y)\,dy = e^{-sb} M(s).$$

A graphical illustration of Eq. (2.5) is given in Fig. 2.8.
Apply Eq. (2.4) to $x_1+\ldots+x_n$ and let $b = na$, we establish

$$\mathrm{Prob}\left[x_1 + \cdots + x_n \geq na\right] \leq e^{-sna}\left[M(s)\right]^n = e^{-n(sa-\log M(s))}, \text{ for all } s\geq 0.$$

This is known as the Chernoff bound (Chernoff 1952). Define a function $R(s)=sa-\log M(s)$. Then we have

$$\mathrm{Prob}\left[x_1 + \cdots + x_n \geq na\right] \leq e^{-nR(s)}, \text{ for all } s \geq 0. \qquad (2.6)$$

We shall use it to establish the exponential decaying rate for $\mathrm{Prob}[x_1+\ldots+x_n\geq na]$. Note although Eq. (2.6) looks already like a bound implying exponential decaying, there is a gap between it and the desired Eq. (2.2) where we need a *positive constant* β. We close the gap by showing that there is an $s^*\geq 0$ such that $R(s^*)>0$. For simplicity, we assume that $\mu = 0$ and $a > 0$ is a constant. (The reader is required to extend the result to the general case where $\mu \neq 0$ below.) Then $\mu = M'(0) = 0$. Consider the Taylor expansion of $M(s)$ around $s = 0$,

$$R(s) = sa-\log(M(0)+M'(0)s+o(s))=sa-o(s).$$

Thus there exists a $s^*>0$ such that

$$R(s^*) = s^*a-\log(M(s^*))>0.$$

We can then choose $\beta = R(s^*)$.

Exercise 2.3: Show that for general μ, as long as $a>\mu$, there exists a positive β such that

$$\mathrm{Prob}[x_1+\ldots+x_n\geq na]\leq e^{-n\beta}.$$

Exercise 2.4: Show that for general μ, as long as $b<\mu$, there exists a positive β such that

$$\mathrm{Prob}[x_1+\ldots+x_n\leq nb]\leq e^{-n\beta}.$$

4.2 Exponential convergence w.r.t. order

With the large deviation theory, given two designs with distinct true performances, it is clear from above (Section 4.1) why order comparison is easy and converges rapidly. The problem is, in general, we have a large number of designs. In this subsection, we argue that the benefit of exponential convergence on order comparison for two designs is preserved for the general situation with N designs. The idea is to upper bound the misalignment probability (the overlap area for the two design case) by the sum of probabilities that sample mean deviates from true performance by the amount δ for each design, where δ is half of the minimal gap Δ between true performances. Here $\Delta = \min_{i=1,\ldots,N-1}\left(J\left(\theta_i\right)-J\left(\theta_{i+1}\right)\right)$ can be viewed as the counterpart of *B-A* for the two-design case.

Given a size s, let S_n be the selected set of size s according to the horse race rule after we obtain the observed (estimated) performance $\widehat{\overline{J}}\left(\theta,n\right)$ based on n replications for all designs θ. Given also a size g of the good enough set G and the alignment level k such that $1 \le k \le \min(g,s)$. Our purpose is to show there exists a positive β such that

$$\text{Prob}\left[\left|S_n \cap G\right| \ge k\right] = 1 - O\left(e^{-n\beta}\right) \tag{2.7}$$

as long as the moment generating function $E[e^{sL(\theta,1)}]$ exists for all $s \in (-d,d)$ for some $d > 0$. It is sufficient to show the result Eq. (2.7) for the case $k = \min(g,s)$ since $\text{Prob}[|S_n \cap G| \ge k] \le \text{Prob}[|S_n \cap G| \ge k']$ for all $k' < k$. The reason is that $\min(g,s)$ is the highest alignment level and increasing the required alignment level always makes alignment harder (lower the alignment probability). Assume that the N designs are indexed such that the true performance value J is sorted in ascending order,

$$J(\theta_1) < J(\theta_2) < J(\theta_3) < \ldots < J(\theta_N).$$

Recall that the observed values $\widehat{\overline{J}}\left(\theta_i,n\right)$ used by the horse race rule are taken as sample mean of performance values of designs, that is $\overline{\widehat{J}}\left(\theta_i,n\right) = \frac{1}{n}\sum_{j=1}^{n}L\left(\theta_i,j\right), i = 1,2,\ldots,N$, with $L(\theta_i,j), j = 1,2,\ldots$ as i.i.d. observations. Without loss of generality, we assume our optimization problem is to find the minimum. Sort the sequence $\overline{\widehat{J}}\left(\theta_i,n\right), i = 1,2,\ldots,N$

in ascending order and denote the design ranking no. i as $\theta_{[i]}$ [5]. In other

words, $\bar{J}\left(\theta_{[1]},n\right) \leq \bar{J}\left(\theta_{[2]},n\right) \leq \ldots \leq \bar{J}\left(\theta_{[N]},n\right)$. So $\theta_{[i]}$, $i=1,2,\ldots,N$ are random variables taking value from the design space $\Theta_N=\{\theta_1,\theta_2,\ldots,\theta_N\}$. The selected set by horse race rule is a random set given as $S_n=\{\theta_{[1]},\theta_{[2]},\ldots,\theta_{[s]}\}$. The alignment probability can be expressed as

$$\text{Prob}\left[\left|S_n \cap G\right| \geq k\right] = \text{Prob}\left[\left|\{\theta_{[1]},\theta_{[2]},\ldots,\theta_{[s]}\} \cap \{\theta_1,\theta_2,\ldots,\theta_g\}\right| \geq k\right].$$

Here g is the size of the good enough set G. To prove $\text{Prob}[|S_n \cap G| \geq k] \geq 1-e^{-n\beta}$ where $k=\min(g,s)$, it is equivalent to show $\text{Prob}[|S_n \cap G| < \min(g,s)] \leq Ce^{-n\beta}$ for some positive constant C. To prove this, denote the minimal gap between any two of true performance values as $\Delta = \min_{i=1,\ldots,N-1}\left(J\left(\theta_i\right) - J\left(\theta_{i+1}\right)\right)$

and introduce two events

$$\text{Event } A = \{|S_n \cap G| < \min(g,s)\}$$
$$\text{Event } B = \{\text{there exists one } \theta_i \text{ in } \Theta_N \text{ s.t. } \left|\bar{J}\left(\theta_i,n\right) - J\left(\theta_i\right)\right| \geq \delta\}$$

where δ is half of the minimal gap Δ. Event A is the misalignment event under level $k = \min(g,s)$, i.e.,

$$\text{Prob}[|S_n \cap G| < \min(g,s)] = \text{Prob}[\text{Event } A].$$

Event B is the event that at least one design's sample mean deviates from its true value over half of the minimal gap Δ. In Fig. 2.9, Event B occurs when at least one design's sample must fall in either dotted area ($\bar{J}\left(\theta_i,n\right) - J\left(\theta_i\right) < -\delta$) or gray area ($\bar{J}\left(\theta_i,n\right) - J\left(\theta_i\right) > \delta$). It is clear from Fig. 2.9 that if *every* design's sample mean stays in the interval centered at the design's true performance value with width Δ (or less) or equivalently the deviation from the true performance is less than $\delta = \Delta/2$, there will be *no* swap in the order of sample means and the alignment level $k = \min(g,s)$ is achieved. This implies that, for a misalignment to occur (furthermore some swaps in sample means to occur), Event B must occur.

---

[5] Please note that $\theta_{[i]}$ depends on n. We assign indices randomly to designs when they tie.

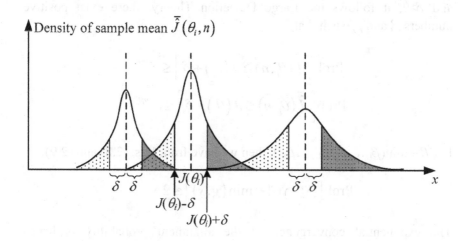

Fig. 2.9. If sample mean stays within δ distance from true performance for each design, no misalignment will happen.

Thus we know that Event A is a subset of Event B and

$$\text{Prob}\Big[|S_n \cap G| < \min(g,s)\Big] = \text{Prob}\big[\text{Event } A\big] \le \text{Prob}\big[\text{Event } B\big]. \quad (2.8)$$

Simple estimation gives

$$\begin{aligned}
\text{Prob}\big[\text{Event } B\big] &= \text{Prob}\left[\bigcup_{i=1,\ldots,N}\Big\{\big|\bar{\tilde{J}}(\theta_i,n)-J(\theta_i)\big|\ge\delta\Big\}\right] \\
&\le \sum_{i=1}^{N}\text{Prob}\Big[\big|\bar{\tilde{J}}(\theta_i,n)-J(\theta_i)\big|\ge\delta\Big] \\
&= \sum_{i=1}^{N}\text{Prob}\Big[\bar{\tilde{J}}(\theta_i,n)\ge J(\theta_i)+\delta\Big] \\
&\quad + \sum_{i=1}^{N}\text{Prob}\Big[\bar{\tilde{J}}(\theta_i,n)\le J(\theta_i)-\delta\Big].
\end{aligned} \qquad (2.9)$$

This is a direct extension of our estimation on misalignment probability for the two-design case (using the sum of gray area and dotted area in Fig. 2.7) to the general case where we have N designs as shown in Fig. 2.9. Now an upper bound for the misalignment probability is given by the sum of N gray areas and N dotted areas associated to the N designs. Since for every design, the true performance J is the mean value of its observed value L

and $\delta > 0$, it follows the Large Deviation Theory, there exist positive numbers β_i and β_i' such that

$$\text{Prob}\left[\widehat{\overline{J}}(\theta_i,n) \geq J(\theta_i)+\delta\right] \leq e^{-n\beta_i}$$

$$\text{Prob}\left[\widehat{\overline{J}}(\theta_i,n) \leq J(\theta_i)-\delta\right] \leq e^{-n\beta_i'}.$$

Let $\beta = \min(\beta_1,\ldots,\beta_N,\beta_1',\ldots\beta_N')$, then we have from Eqs. (2.8) and (2.9)

$$\text{Prob}\left[|S_n \cap G| < \min(g,s)\right] \leq 2Ne^{-n\beta}.$$

The exponential convergence of the alignment probability is hence established by noting the size N of design space is fixed.

So far, we have shown the exponential convergence of OO w.r.t. order using the large deviation theory and estimation on misalignment probability for designs with i.i.d. observations. The exponential convergence of the alignment probability can be generalized to the situation of regenerative simulation[6], where performances are estimated by taking time average over a single sample path based on the ergodic properties of discrete event systems. When we carry out a simulation of length t and obtain some observed value $\widehat{\overline{J}}(\theta,t)$ for all designs θ, by applying horse race rule, we can decide a selected set S_t. Since S_t depends on t, which is a measure of computation budget, the question now becomes where $\text{Prob}[|S_t \cap G| \geq k]$ converge exponentially as $t \to \infty$? The exponential convergence for this case means that there exists $\beta > 0$ such that

$$\text{Prob}\left[|S_t \cap G| \geq k\right] = 1 - O\left(e^{-t\beta}\right). \tag{2.10}$$

It was proved in (Xie 1997) that when the Heidelberger and Meketon's esmtiators (Heidelberger and Meketon 1980) defined in Eq. (2.11) or time average estimators defined in Eq. (2.12) below are used as observed value $\widehat{\overline{J}}(\theta,n)$, under mild condition, there must exist a $\beta > 0$ such that Eq. (2.10) holds.

---

[6] The basic idea of the approach of regenerative simulation is that a stochastic process may be characterized by random points in time when it "regenerative" itself and become independent of its past history. (See also Appendix A.)

Intuitively, since a regenerative simulation is equivalent to many periods of statistically independent replications of the system sample path, the validity of Eq. (2.10) is totally reasonable. The proofs can be found in (Xie 1997) and will be omitted here.

Note, this type of results were first obtained in (Dai 1996) showing that the best observed design is indeed a "good" design. (Xie 1997) extended these results to the general setting we describe here. Extensions of (Dai 1996) to the situation using common random variables in simulation have been made in (Dai and Chen 1997).

To define the estimators and the exponential convergence mathematically, we will introduce some notations. Let $T_i(\theta)$ be the i-th regeneration epoch, $i = 0,1,2,\ldots$, where $T_0(\theta)$ is the initial delay. Let $\tau_i(\theta)$ be the length of i-th regeneration cycle, $i = 0,1,2,\ldots$. Then $T_i(\theta) = \sum_{j=0}^{i} \tau_j(\theta)$. Suppose the interested performance value on sample path at time t is $L_t(\theta)$ with $|L_t(\theta)| \leq C$ for some constant C. Let $L(\theta, i) = \int_{s=T_{i-1}(\theta)}^{T_i(\theta)} L_s(\theta) ds$ be the total sample performance in the i-th regeneration cycle.

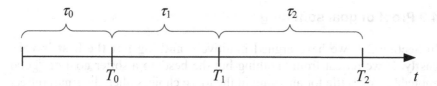

Fig. 2.10. An illustration for the regeneration cycles

Let $K(\theta, t)$ be the number of regeneration cycles completed by time t. Then the Heidelberger and Meketon's estimator is defined as

$$\bar{\tilde{J}}(\theta, t) = \frac{\sum_{i=1}^{K(\theta,t)+1} L(\theta, i)}{\sum_{i=1}^{K(\theta,t)+1} \tau_i(\theta)}, \tag{2.11}$$

and the time average estimator is defined as

$$\overline{\tilde{J}}\left(\theta,t\right)=\frac{1}{t}\int\limits_{s=0}^{t}L_{s}\left(\theta\right)ds. \tag{2.12}$$

We assume that the regeneration process has i.i.d cycles, i.e., $\{(\tau_i(\theta),$ $L(\theta,i)),$ $i=1,2,...\}$ is a sequence of i.i.d. random variables. Denote $m(s,\theta)=E[e^{s\tau_0(\theta)}]$ as the mgf (moment generating function) of the initial delay $\tau_0(\theta)$ and $M(s,\theta)=E[e^{s\tau_1(\theta)}]$ as the mgf of the length $\tau_1(\theta)$ of the first regeneration cycle $\tau_1(\theta)$. A sufficient condition for Eq. (2.10) to hold for the estimators in Eq. (2.11) or Eq. (2.12) is that $m(s,\theta)$ and $M(s,\theta)$ exist for all $s\in(-\delta,\delta)$ for some $\delta>0$. Note that this existence of a finite mgf was later shown by Fu and Jin to be both a necessary and sufficient condition in (Fu and Jin 2001). They have also shown how one can recover the exponential convergence rate in cases where the mfg is not finite. (Well-known distributions that do not possess a finite mgf include the lognormal distribution and certain gamma distributions.) In particular, by working with appropriately truncated versions of the original random variables, the exponential convergence can be recovered.

4.3 Proof of goal softening

In Section 2.1, we have argued intuitively **nothing but the best is very costly.** If we retreat from "nothing but the best" to a softer goal of "good enough", e.g., settle for anything in the top-g choices, then the small retreat can buy us quite a bit in the ease of the computational burden. In this subsection, we make a rigorous justification for this point and will show that the alignment probability for both blind pick selection rule and horse race selection rule converges exponentially to 1, as the size g of the good enough set and the size s of the selected set increase.

4.3.1 Blind pick

First, let us show the exponential convergence result for blind pick. It will be used as a base for proving the exponential convergence of the horse race rule. In fact, we will prove that the alignment probability of blind pick rule is always a lower bound for the alignment probability of horse race rule. This is reasonable, since no knowledge is used in BP, and in HR some, though imperfect, knowledge is used to select the set S. Let N be the size of the design space, g and s be the size of good enough set G and the size of selected set S respectively. For blind pick, the misalignment probability

Prob[|S∩G|=0] is given by (see full derivation in Eq. (2.37) in Section 5.1 below)

$$\frac{\binom{N-g}{s}}{\binom{N}{s}},$$

where $\binom{N}{s}$ represents the number of different choices of s designs out of N distinguished ones, i.e.,

$$\binom{N}{s} = \frac{N!}{s!(N-s)!}.$$

Thus, the alignment probability Prob[|S∩G|≥1] is given by

$$\text{Prob}\Big[|S \cap G| \geq 1\Big] = 1 - \text{Prob}\Big[|S \cap G| = 0\Big] = 1 - \frac{\binom{N-g}{s}}{\binom{N}{s}} \qquad (2.13)$$

$$= 1 - \frac{\dfrac{(N-g)!}{s!(N-g-s)!}}{\dfrac{N!}{s!(N-s)!}} = 1 - \frac{(N-g)(N-g-1)\cdots(N-g-s+1)}{(N)(N-1)\cdots(N-s+1)}. \qquad (2.14)$$

Since $\dfrac{N-g-i}{N-i} \leq \dfrac{N-g}{N} = 1 - \dfrac{g}{N}$ for all $i = 0,1\ldots,s-1$, we have

$$\text{Prob}\Big[|S \cap G| \geq 1\Big] \geq 1 - \Big(1 - \frac{g}{N}\Big)^s. \qquad (2.15)$$

Furthermore, since $1-x \leq e^{-x}$ holds for all x, we can bound Prob[|S∩G|≥1] from below as

$$\text{Prob}\Big[|S \cap G| \geq 1\Big] \geq 1 - e^{-\frac{gs}{N}}, \qquad (2.16)$$

which is the desired result, i.e., alignment probability for blind pick converges exponentially w.r.t. the size of the set G and S.

Exercise 2.5: Draw a figure to verify the above statement that $1 - x \le e^{-x}$.

4.3.2 Horse race

Now we are going to present the convergence result for the horse race selection rule alignment probability. We assume that the i.i.d. noise $w(\theta_i, n) = W_i(n)$, $i=1,2,\ldots,N$, has the common cumulative continuous distribution function $F_n(x)$ and density function $f_n(x)$ and has zero mean.

With the help of the relation

$$\text{Prob}\left[|\, S \cap G\,| \ge 1\right] = 1 - \text{Prob}\left[|\, S \cap G\,| = 0\right],$$

we will show that for horse race selection rule, the alignment probability $\text{Prob}\left[|\, S \cap G\,| \ge 1\right]$ is bounded from below by the function $1 - e^{-\frac{gs}{N}}$, that is

$$\text{Prob}\left[|\, S \cap G\,| \ge 1\right] \ge 1 - e^{-\frac{gs}{N}}. \tag{2.17}$$

This is quite reasonable, since BP utilizes no knowledge of the problem, while the S selected by the horse race rule can only improve upon the AP. We can expect the same exponential convergence. While it is intuitively reasonable to suppose that any crude model for picking the set S must result in better performance than blind pick, it is nevertheless important to rule out crude models that may appear sensible on the surface but actually favor bad designs unknowingly. Consequently, we must prove that the S obtained based on a horse race model will indeed perform better and results in better AP than on a blind pick model. This is the purpose of this section.

To establish exponential convergence for horse race by leveraging the results in Section 4.3.1 on AP for BP, we should follow two steps:

Step 1. Identify Least Favorable Configuration (LFC) for horse race misalignment probability.

Step 2. Evaluate misalignment probability under LFC and prove its equivalence to that of blind pick.

Least Favorable Configuration (LFC) is well known in Ranking and Selection literature (Barr and Rizvi 1966). The general idea is to find and take advantage of some monotone properties in a set of distributions with a parameter (such as mean value) to specific setting of the parameter under which certain ranking or selection probability of interest achieves maximum or minimum. In our case, we should use true performance as the parameter and aim at finding LFC for misalignment probability under the horse race selection rule. Let us first take a close look at the misalignment event under horse race.

For a direct derivation of the result, the basic idea is this: whenever the observed performance of every design in G is no better than that of at least s designs not in G, none of the designs in G will be selected in which case [$|S \cap G|=0$]. Fig. 2.11 shows a case of $N = 6$ designs with $g = 2$ and $s = 3$. In the figure, we show the procedure of generation of observation performance $\hat{\bar{J}}(\theta_i, n) = J(\theta_i) + W_i(n)$ from the true performance $J(\theta_i)$ by adding noise $W_i(n)$, $i = 1, 2, \ldots, 6$. The best observed performance of the two good enough designs (black balls in the lower part of the figure) is indicated by value A. Since it is greater than the observed performance of the three designs not in G (white balls), the select set by horse race contains only white balls which means a misalignment occurs. In the figure, we order the observed performance of all four designs not in G and indicate the s-th (third) value as B. We observe that $B<A$ is true, when misalignment happens.

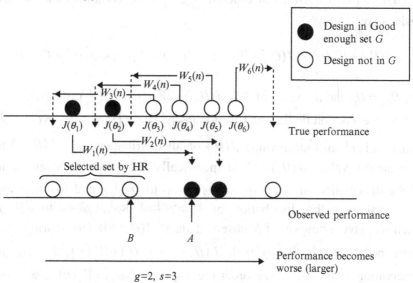

Fig. 2.11. An illustration for the misalignment event under horse race rule

To characterize $\text{Prob}\left[|S \cap G| = 0\right]$ in detail, we divide a given set of observation data $\bar{J}\left(\theta_i, n\right) = J\left(\theta_i\right) + W_i\left(n\right)$ into two groups, the observed data for good designs $G = \{\theta_1, \ldots \theta_g\}$ and the data for bad designs $(\theta_{g+1}, \ldots \theta_N)$. We order the N-g observation data for bad designs such that

$$J\left(\theta_{[g+1]}\right) + W_{[g+1]}\left(n\right) < \ldots < J\left(\theta_{[N]}\right) + W_{[N]}\left(n\right). \qquad (2.18)$$

For a misalignment to happen when using horse race, we observe that, there must be at least s "bad" designs outperform good designs $\theta_1, \ldots, \theta_g$. Or put it in another way, all observed performances for good designs must be larger than $B = J\left(\theta_{[g+s]}\right) + W_{[g+s]}(n)$. Denote the best observed performance of good designs as $A = \min_{j \in \{1, \ldots, g\}}\left(J(\theta_j) + W_j(n)\right)$. Then a misalignment simply means $B < A$ holds.

$$\text{Prob}\left[|S \cap G| = 0\right] = \text{Prob}\left[B < A\right]. \qquad (2.19)$$

For example, in Fig. 2.11, the value of A is $\min(J(\theta_1) + W_1(n), J(\theta_2) + W_2(n)) = J(\theta_1) + W_1(n)$, the ordered N-g = 4 observation data for bad designs are

$$J(\theta_3) + W_3(n) < J(\theta_4) + W_4(n) < J(\theta_5) + W_5(n) < J(\theta_6) + W_6(n).$$

So, $\theta_{[5]} = \theta_5$ and the value of B is $J(\theta_{[5]}) + W_{[5]}(n) = J(\theta_5) + W_5(n)$.

Now we work on finding the LFC for $\text{Prob}\left[B < A\right] = \text{Prob}\left[|S \cap G| = 0\right]$. Our idea is to shift mean value $J(\theta_i)$ of all distributions of $\bar{J}\left(\theta_i, n\right)$ to a common value $J(\theta_g)$. More specifically, we move the mean value of the distribution of every good design up to $J(\theta_g)$ and to move the mean value of the distribution of every bad design *down* to $J(\theta_g)$. Then we have a new set of N observed data $J(\theta_g) + W_i(n)$ sharing the same noise sample $W_i(n)$ with $\bar{J}\left(\theta_i, n\right) = J\left(\theta_i\right) + W_i\left(n\right)$, the original observation with noise. We order the N-g data $J(\theta_g) + W_i(n)$ associated with bad designs such that

$$J\left(\theta_g\right)+W_{[g+1]}(n)<\ldots<J\left(\theta_g\right)+W_{[N]}(n). \tag{2.20}$$

Note since the mean values are reduced for the data of bad designs, the value $J(\theta_g)+W_{[g+s]}(n)$ appearing in the ordered sequence in Eq. (2.20) must be no greater than its counterpart $B=J(\theta_{[g+s]})+W_{[g+s]}(n)$ appeared in Eq. (2.18). At the same time, as the counterpart of A, the random variable $\min_{j\in\{1,\ldots,g\}}(J(\theta_g)+W_j(n))$ must be no less than A. Let us denote $A'=\min_{j\in\{1,\ldots,g\}}(J(\theta_g)+W_j(n))$ and $B'=J(\theta_g)+W_{[g+s]}(n)$. Fig. 2.12 shows this procedure of finding LFC for the designs in Fig. 2.11. Recall, we have $g=2$ and $s=3$. We move the mean value of the observed performance's distribution to $J(\theta_2)$ but keep the sample of noise $W_i(n)$ the same as in Fig. 2.11. The new observation data become $J(\theta_2)+W_i(n)$, $i=1,2,\ldots,6$, and their values are as shown in the line "observed performance (LFC)". For reader's convenience, we also show original observation data in the bottom of the figure. The best value for good designs under new observation i.e., A', equals $\min(J(\theta_2)+W_1(n),J(\theta_2)+W_2(n))=J(\theta_2)+W_2(n)$. We order the new observations of the four bad designs as $J(\theta_2)+W_3(n)<J(\theta_2)+W_4(n)<J(\theta_2)+W_5(n)<J(\theta_2)+W_6(n)$. So, the third ($s$-th) value in this sequence is $J(\theta_2)+W_5(n)$, $\theta_{(5)}=\theta_5$ and the value of B' is $J(\theta_2)+W_5(n)$. Two important observations from this example are $B'<B$ and $A<A'$, as a result of our way of generating new observations.

With the new distributions defined above, we are able to establish an upper bound for $\text{Prob}[B<A]$ in Eq. (2.19), that is

$$\text{Prob}[B<A]\le\text{Prob}[B'<A']. \tag{2.21}$$

In fact, we have noted that $B'<B$ and $A<A'$. As a result, $B<A$ always implies $B'<A'$. This shows indeed the shift we made leads to the LFC for the misalignment probability.

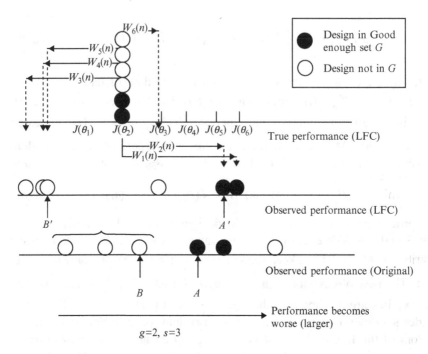

Fig. 2.12. An illustration of the Least Favorable Configuration for designs in Fig. 2.11

Exercise 2.6: Show Eq. (2.21) is true.

Eq. (2.21) is the LFC result in our context. Its advantage is, now instead of dealing with the observation data $\hat{\bar{J}}(\theta_i, n) = J(\theta_i) + W_i(n)$ which are all following different distributions with different means, we need only to deal with the case of i.i.d. observations $W_j(n)$ plus a constant.

In summary, Fig. 2.12 and our arguments above show, for horse race selection rule, misalignment occurs when the best value (A) of good designs is greater (for minimization problem) than the values of s bad designs (we denote B as their maximum) in the observation data. Although evaluating the probability of $B < A$ is generally difficult due to the heterogeneous nature of distributions generating observations of designs, this characterization of misalignment enables us to find the new setting that provides a tractable upper bound for misalignment probability, which turns out to be the same as blind pick. The new setting is tractable since observations of all designs obey i.i.d. distributions. The new setting provides an upper bound on misalignment probability (thus it is a LFC) because the gap between the maximum of s bad designs B' and its best observed value for good designs A' is always greater than that of B and A. The connection between the new setting and blind pick is natural because independent draw of samples from the same distribution gives no preference on any specific design and all designs have equal chance to be selected which is the same as blind pick.

Now we proceed to evaluate misalignment probability under the LFC, or equivalently to calculate $\text{Prob}\left[B' < A'\right]$. This is the second step in order to establish the exponential convergence for horse race. It will be shown below that $\text{Prob}\left[B' < A'\right] = \binom{N-g}{s}\binom{N}{s}^{-1}$ which is exactly the misalignment probability already given in Section 4.3.1 for blind pick. For readers not interested in the mathematical details, you can go directly to the end of this section (the texts below Eq. (2.36)).

Without loss of generality, we can simplify the analysis of misalignment probability of our LFC by ignoring the common constant $J(\theta_g)$ in all observations $J(\theta_g) + W_i(n)$ and directly define $A^* = \min_{j \in \{1,\dots,g\}} W_j(n)$ and $B^* = W_{[g+s]}(n)$ based on the zero mean sample $W_i(n)$. Denote their densities and distributions as $\phi_{A^*}(x), \phi_{B^*}(y)$ and $\Phi_{A^*}(x), \Phi_{B^*}(y)$ respectively. Then we can write

$$\text{Prob}\left[B' < A'\right] = \text{Prob}\left[B^* < A^*\right] = \int_{-\infty}^{+\infty} \text{Prob}\left[B^* < x\right]\phi_{A^*}(x)\,dx$$

$$= \int_{-\infty}^{+\infty} \int_{-\infty}^{x} \phi_{B^*}(y)\,dy\,\phi_{A^*}(x)\,dx \tag{2.22}$$

based on the fact that $A^* = \min_{j \in \{1,\dots,g\}} W_j(n)$ and $B^* = W_{[g+s]}(n)$ are independent.

Exercise 2.7: Verify $\text{Prob}\left[B' < A'\right] = \text{Prob}\left[B^* < A^*\right]$.

Hint: Compare the samples $J(\theta_g) + W_i(n)$ and $W_i(n)$ $i = 1, 2, \dots, N$.

Exercise 2.8: Show

$$\phi_{A^*}(x) = g\left[1 - F_n(x)\right]^{g-1} f_n(x). \tag{2.23}$$

Hint: use Eq. (2.24) below.

In order to evaluate $\phi_{A^*}(x)$, we first decide the distribution function $\Phi_{A^*}(x)$. It is straightforward to see that

$$
\begin{aligned}
\Phi_{A^*}(x) &= \text{Prob}\left[A^* < x\right] = 1 - \text{Prob}\left[\min_{j \in \{1,\dots,g\}} W_j(n) > x\right] \\
&= 1 - \prod_{j=1}^{g} \text{Prob}\left[W_j(n) > x\right] = 1 - \left[1 - F_n(x)\right]^g .
\end{aligned}
\tag{2.24}
$$

To evaluate $\phi_{B^*}(y)$, we first find its distribution $\Phi_{B^*}(y)$. For a given value y, when $B^* = W_{[g+s]} < y$ is true, the bad design data $W_{g+1}(n), \dots,$ $W_N(n)$ are divided further into two groups, one group Λ contains m ($\geq s$) values which are less than A^*, the other group Λ^- contains all the remaining N-g-m bad designs which are greater than y. According to the value of m and the way two groups formed, we can express the distribution $\Phi_{B^*}(y)$ as the following exclusive unions:

$$
\begin{aligned}
\Phi_{B^*}(y) &= \text{Prob}\left[W_{[g+s]} < y\right] \\
&= \text{Prob}\left[\bigcup_{\substack{\Lambda \subset \{g+1,\dots,N\} \\ \Lambda \text{ has at least } s \text{ elements}}} \left\{\max_{i \in \Lambda} W_i(n) < y \text{ and } \min_{j \in \Lambda^-} W_j(n) > y\right\}\right]
\end{aligned}
\tag{2.25}
$$

$$
= \sum_{m=s}^{N-g} \sum_{\substack{\Lambda \subset \{g+1,\dots,N\} \\ \Lambda \text{ has } m \text{ elements}}} \text{Prob}\left[\max_{i \in \Lambda} W_i(n) < y \text{ and } \min_{j \in \Lambda^-} W_j(n) > y\right] .
\tag{2.26}
$$

We have

$$
\text{Prob}\left[\max_{i \in \Lambda} W_i(n) < y\right] = \left[F_n(y)\right]^m
\tag{2.27}
$$

and

$$
\text{Prob}\left[\min_{j \in \Lambda^-} W_j(n) > y\right] = \left[1 - F_n(y)\right]^{N-g-m} .
\tag{2.28}
$$

Notice that $\max_{i\in\Lambda} W_i(n)$ and $\min_{j\in\Lambda^-} W_j(n)$ are independent, we have from Eqs. (2.26)–(2.28) that

$$\Phi_{B*}(y) = \sum_{m=s}^{N-g} \sum_{\substack{\Lambda\subset\{g+1,\dots,N\} \\ \Lambda \text{ has } m \text{ elements}}} [F_n(y)]^m [1-F_n(y)]^{N-g-m}$$

$$= \sum_{m=s}^{N-g} \binom{N-g}{m} [F_n(y)]^m [1-F_n(y)]^{N-g-m}.$$

(2.29)

Taking derivative on Eq. (2.29) yields

$$\phi_{B*}(y) = s\binom{N-g}{s} [F_n(y)]^{s-1} [1-F_n(y)]^{N-g-s} f_n(y). \qquad (2.30)$$

Exercise 2.9: Show Eq. (2.30).

Now with expressions Eqs. (2.23) and (2.30) of $\phi_{A*}(x)$ and $\phi_{B*}(y)$ plugged in, we are ready to proceed on the integration in Eq. (2.22). We have

$$\text{Prob}[B' < A']$$

$$= \int_{-\infty}^{+\infty} \int_{-\infty}^{x} s\binom{N-g}{s} [F_n(y)]^{s-1} [1-F_n(y)]^{N-g-s} f_n(y)dy g[1-F_n(x)]^{g-1} f_n(x)dx.$$

(2.31)

Using substitution, letting $u = F_n(x)$ and $v = F_n(y)$, we can further simplify Eq. (2.31)

$$\text{Prob}[B' < A'] = gs\binom{N-g}{s} \int_0^1 \int_0^u v^{s-1} [1-v]^{N-g-s} \, dv [1-u]^{g-1} \, du. \qquad (2.32)$$

Using induction method, one can show that

$$gs\int_0^1 \int_0^u v^{s-1} [1-v]^{N-g-s} \, dv [1-u]^{g-1} \, du = \binom{N}{s}^{-1}. \qquad (2.33)$$

As a consequence, we have

$$\text{Prob}\left[B' < A'\right] = \frac{\binom{N-g}{s}}{\binom{N}{s}}, \tag{2.34}$$

which implies

$$\text{Prob}\left[\left|S \cap G\right| = 0\right] = \frac{\binom{N-g}{s}}{\binom{N}{s}} \tag{2.35}$$

and furthermore

$$\text{Prob}\left[\left|S \cap G\right| \geq 1\right] = 1 - \text{Prob}\left[\left|S \cap G\right| = 0\right] = 1 - \frac{\binom{N-g}{s}}{\binom{N}{s}}. \tag{2.36}$$

This shows that for the worst case where all observed performance values $\bar{\hat{J}}(\theta_i, n)$ are i.i.d., the alignment probability of horse race rule is the same as that of blind pick. As a result, for general cases where the true performance values *are* different, the alignment probability of horse race rule is bounded from below by the blind pick alignment probability

$$1 - \frac{\binom{N-g}{s}}{\binom{N}{s}}.$$

The desired result Eq. (2.17) then follows from the exponential convergence w.r.t. the size of G and S for blind pick.

Note exponential convergence of OO w.r.t. the size of G and S was originally given in (Lee et al. 1999). The noises were assumed to obey normal distributions. This assumption allows one to find the same LFC for noises with non-identical distributions, but as we have shown above, the normal distribution assumption is not necessary for our case where noises obey identical distribution.

Exercise 2.10: If in the observation data $\overset{\wedge}{\tilde{J}}\left(\theta_i, n\right) = J(\theta_i) + W_i(n)$, the noises $W_i(n)$ obey normal distributions $N(0, \sigma_i^2 / n)$. Show the misalignment probability $\text{Prob}\left[| S \cap G | = 0\right]$ is no greater than the scenario where the noises $W_i(n)$ obey normal distributions $N(0, \sigma^2 / n)$ with $\sigma^2 = \max_{i=1,\dots,N} \sigma_i^2$.

5 Universal alignment probabilities

The discussion in Section 2 and the demonstration in Section 3 suggest that the concept of alignment probability $\text{Prob}[|G \cap S| \geq k]$ is rather general and can be problem independent. Thus it is possible to establish some universal scheme for all optimization problems to help narrow down the search for "good enough" designs as a function of the number of crude samples taken, N, the approximate size of the estimation error, σ^2, the type of problem class, C, and finally the selection procedure used. This can be very useful during the initial phase in many problems that involve (i) a structureless and immense search space and (ii) performance evaluation that is corrupted by large noise/error and/or is computationally intensive. We explore this possibility below (see also (Lau and Ho 1997)). Alert reader may point out here that our aim here bears resemblance to the extensive literature in statistics on rank and selection (R&S) (Gupta and Panchapakesan 1979; Santer and Tamhane 1984; Goldman and Nelson 1994). There are, however, two major differences. First, the R&S schemes deal with a search space of usually less than a hundred[7], often in tens (such as in comparison study of the efficacy of different drugs) while we consider subset selection from Θ that has size in billions and zillions. Second, the cardinal notions of "distance of the best from the rest" and the probability of "coincidence of the observed best with the true best" used in R&S have very little significance in our problem domain. Instead, we focus on softened criterion and different selection procedures.

---

[7] Although recent development of R&S allows to deal with a design space as large as 500 (Nelson et al. 2001; Wilson 2001), this is still comparatively small than the size that OO can handle.

5.1 Blind pick selection rule

We obtain the simplest result on alignment probability by using the blind pick selection rule, i.e., we blindly pick out the members from the selected set, S without any evaluation of the performances from N samples. Equivalently, we can say that the performances in Eq. (2.1) is sampled with the noise variance being infinite (in the demonstration of Section 3, we used the noise distribution of $U[0,10000]$ to approximate the blind pick selection rule when the range of the true performance is $[0,200]$). For given size of S and G being s and g respectively, the alignment probability for blind pick (BP) is

$$AP(s,g,k,N/BP) = \sum_{i=k}^{\min(g,s)} \frac{\binom{g}{i}\binom{N-g}{s-i}}{\binom{N}{s}} = \text{Prob}\left[|G \cap S| \geq k\right]. \quad (2.37)$$

Exercise 2.11: Try to derive Eq. (2.37) before reading the explanation below.

The validity of Eq. (2.37) can be seen as follows: There are total $\binom{N}{s}$[8] ways of picking s out of N designs. Suppose i of these s designs actually belong to G, then there are $\binom{g}{i}$ ways for which this is possible. The remaining s-i designs can be distributed in $\binom{N-g}{s-i}$ ways. The product of these two factors constitutes the total number of ways that we find exact i members of G by picking out s designs out of N. Dividing this product by $\binom{N}{s}$ yields the probability for i. Summing over all $i \geq k$ gives Eq. (2.37).

---

[8] $\binom{N}{s}$ represents the number of different choices of s designs out of N distinguished ones, i.e., $\binom{N}{s} = \dfrac{N!}{s!(N-s)!}$.

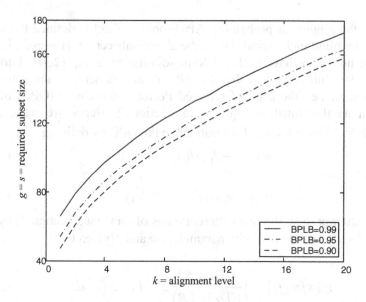

Fig. 2.13. Required subset size vs. alignment level for different alignment probabilities

Fig. 2.13 shows a plot for required size of *s*, with *g* = *s* as a function of alignment level *k* for AP = 0.99, 0.95, and 0.90. These curves can be used as a lower bound (LB) for the UAP for any problems. It is instructive to see that to insure with probability (w.p. for short) 0.99 that there are at least two top 8% choices out of 1000 choices, we only need to blindly pick 80 samples – a more than ten fold reduction in search effort. Note this number *s* is an upper bound for selection since it is done **without any knowledge**. Imagine how much better we can do with some approximate knowledge about the problem. *This is the essence of ordinal optimization!* The next subsection will discuss the first of such less random selection rules.

5.2 Horse race selection rule

In Section 3 we demonstrated the horse race selection rule for *S*. The procedure of this rule is:

- We take *N* samples uniformly from Θ
- Using a crude model, we estimate the performances of these *N* samples as $\hat{J}(\theta_1),...,\hat{J}(\theta_N)$
- We sort these samples according their estimated performances as $\hat{J}(\theta_{[1]}), ... ,\hat{J}(\theta_{[N]})$
- Select the observed top-*s* members of the *N* samples as the selected set *S*.

Then the alignment probability AP≡Prob[|G∩S|≥k] is defined the same way as the blind pick probability in the above subsection. However, in this case we no longer have a closed form solution as in Eq. (2.37). Furthermore, it is intuitively clear that the AP will also depend on the nature of the problem, i.e., the class of Ordered Performance Curve (OPC) of the problem as illustrated in Fig. 2.3 in Section 2. Hence we write AP= $\mathcal{F}(g,s,k,N,C$/Horse Race). If we normalize the OPC by defining

$$y_i = (J_{[i]} - J_{[1]})/(J_{[N]} - J_{[1]}) \tag{2.38}$$

$$x(\theta_{[i]}) \equiv x_i = (i-1)/(N-1) \tag{2.39}$$

we can attempt to fit the five different types of OPC (see Section 3) by the Incomplete Beta Function with parameters α and β given by

$$F(x/\alpha,\beta) \equiv \int_0^x \frac{\Gamma(\alpha+\beta)}{\Gamma(\alpha)\Gamma(\beta)} z^{\alpha-1}(1-z)^{\beta-1} dz \tag{2.40}$$

with the normalized OPC as

$$\Lambda(x/\alpha,\beta) \equiv F\left(x \middle/ \frac{1}{\alpha}, \frac{1}{\beta}\right). \tag{2.41}$$

For different values of α, β we can describe the different shapes of five different types of OPC and their significances in Fig. 2.14 and Fig. 2.15 below, where in Fig. 2.14 the Normalized Performance Densities are the derivatives of the inverse functions of the Normalized OPCs, respectively.

For a given pair of α and β, we can determine the AP by a simple simulation model in the same way as the Excel demo example outlined in Section 3. Extensive simulation has been done on these normalized OPCs (Lau and Ho 1997).

The principal utility of AP in practice is to determine the required size of the set S. For a given problem, the designer/optimizer picks the crude but computationally easy model to estimate the performance. S/he specifies what is meant by "good enough", namely the size of G, g. S/he also have some rough idea of the parameters, σ^2 and C [9] (hence the values α and β in Eq. (2.41) above). For practicality we set AP≥0.95. Then we can

---

[9] A rough idea of C can be gleamed always from the N samples $\hat{J}(\theta_1),\ldots,\hat{J}(\theta_N)$.

experimentally determine a function $Z(g,k/N,C,\sigma^2,AP)$ which tells how large must the set S be in order to insure that it contains at least k member of the set G with high probability. The significance of this information is obvious. We have engineered a reduction of the search and computational effort from Θ to N to $|S| = s$.

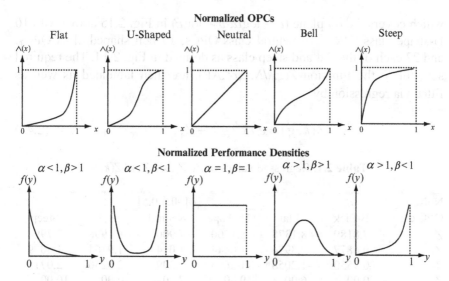

Fig. 2.14. Examples of beta density and corresponding standardized OPCs

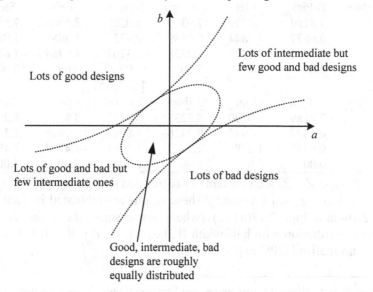

Fig. 2.15. Partitions of the *ab*-plane for five OPC categories, where $a = \log\alpha$, $b = \log\beta$

Extensive simulation experiments have been carried out with a total of 88 normalized OPCs (classified into 5 types of OPCs) using different α and β's

$$\alpha, \beta \in \{\, 0.15, 0.25, 0.4, 0.65, 1.0, 1,5, 2.0, 3.0, 4.5, 8.0 \}$$

which covers the *ab* plane ($a = \log\alpha$, $b = \log\beta$) in Fig. 2.15 above with 10 U-shape class OPCs, 19 neutral class OPCs, 15 bell shaped class OPCs, and 22 each of the flat and steep class as defined in Fig. 2.14. The required sizes of S, the function $Z(g, k/N, C, \sigma^2, AP)$, are then tabulated as well as fitted via regression by

$$Z(k, g) = e^{Z_1} k^{Z_2} g^{Z_3} + Z_4 , \qquad (2.42)$$

Table 2.1. Regressed values of Z_1, Z_2, Z_3, Z_4 in $Z(k,g)$

Noise	∞	$U[-0.5, 0.5]$				
OPC class	B-Pick	Flat	U-shape	Neutral	Bell	Steep
Z_1	7.8189	8.1378	8.1200	7.9000	8.1998	7.7998
Z_2	0.6877	0.8974	1.0044	1.0144	1.9164	1.5099
Z_3	−0.9550	−1.2058	−1.3695	−1.3995	−2.0250	−2.0719
Z_4	0.00	6.00	9.00	7.00	10.00	10.00
Noise	∞	$U[-1.0, 1.0]$				
OPC class	B-Pick	Flat	U-shape	Neutral	Bell	Steep
Z_1	7.8189	8.4299	7.9399	8.0200	8.5988	7.5966
Z_2	0.6877	0.7844	0.8989	0.9554	1.4089	1.9801
Z_3	−0.9550	−1.1795	−1.2358	−1.3167	−1.6789	−1.8884
Z_4	0.00	2.00	7.00	10.00	9.00	10.00
Noise	∞	$U[-2.5, 2.5]$				
OPC class	B-Pick	Flat	U-shape	Neutral	Bell	Steep
Z_1	7.8189	8.5200	8.2232	8.4832	8.8697	8.2995
Z_2	0.6877	0.8944	0.9426	1.0207	1.1489	1.3777
Z_3	−0.9550	−1.2286	−1.2677	−1.3761	−1.4734	−1.4986
Z_4	0.00	5.00	6.00	6.00	7.00	8.00

where Z_1, Z_2, Z_3, Z_4 are constants of regression depending on OPC types, the noise level, g, and k values[10]. These results are tabulated in Table 2.1, Fig. 2.16 and Fig. 2.17(a)–(e) (where we assume the noise contains uniform distribution with half-width W, i.e., $U[-W, W]$, $W = 0.5, 1.0, 2.5$.) against normalized OPC in [0,1].

---

[10] We do not believe a linear regression function would fit the data well. Thus, a product form is the next simplest nonlinear function we can try.

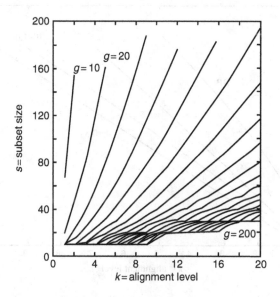

Fig. 2.16. Subset size interpolated from simulated data for the neutral class OPC and $W = 1.0$

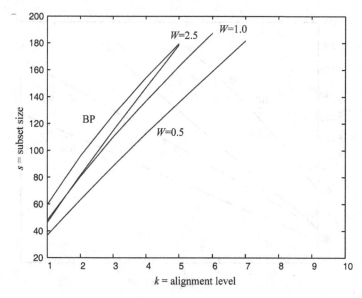

Fig. 2.17(a). Subset size for the flat OPC class at different noise levels with $g = 50$

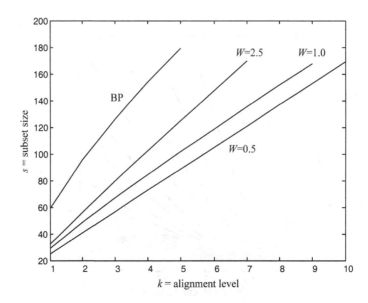

Fig. 2.17(b). Subset size for U-shaped OPC class at different noise levels with $g = 50$

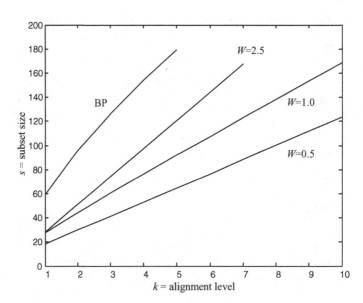

Fig. 2.17(c). Subset size for the neutral OPC class at different noise level with $g = 50$

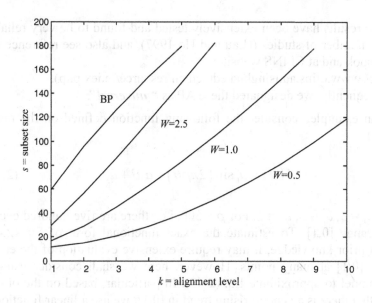

Fig. 2.17(d). Subset size for the bell-shaped OPC class at different noise levels with $g = 50$

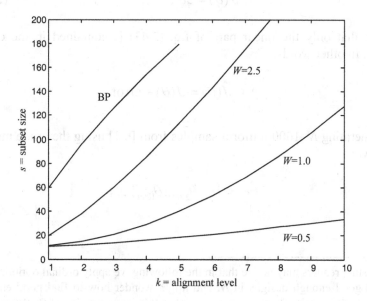

Fig. 2.17(e). Subset size for the steep OPC class at different noise levels with $g = 50$

These results have been extensively tested and found to be very reliable in large number of studies ((Lau and Ho 1997), and also see reference list in this book and at CFINS website:

http://www.cfins.au.tsinghua.edu.cn/en/resource/index.php).

Consequently, we designated these AP as ***"universal"***.

As an example, consider the following function defined on the range $\Theta=[0,1]$

$$J(\theta) = a_1 \sin(2\pi\rho\theta) + a_2\theta + a_3, \tag{2.43}$$

where $a_1 = 3$, $a_2 = 5$, $a_3 = 2$. For $\rho = 500$, i.e., there are five hundred cycles in the range $[0,1]$. To estimate the exact functional form of Eq. (2.43) without prior knowledge, it may require extensive evaluation of the entire domain $[0,1]$ at many points. However, here we shall consider using a crude model to approximate Eq. (2.43). In particular, based on the observation that there is a general rising trend in $[0,1]$, we use a linear function

$$\hat{J}(\theta) = 5\theta. \tag{2.44}$$

Notice that only the linear part of Eq. (2.43) is contained in the crude model. In other words,

$$\hat{J}(\theta) = J(\theta) + \text{error}.$$

By generating $N=1000$ uniform samples from $[0,1]$ using the crude model, we have

$$\hat{\Theta}_N = \{\theta_1, \theta_2, ..., \theta_{1000}\}.\text{[11]}$$

---

[11] Astute readers may notice that in the following we apply ordinal optimization to find good enough designs in N, and might wonder how to find good enough designs in Θ instead. The quick answer is that N is representative of Θ. When both N and Θ are large enough (which this example satisfies) the selected set that is selected from N also has a high probability to contain good enough designs in Θ, and the difference between the two alignment probabilities can be ignored for engineering purpose. But this notion will be quantified and made precise in Chapter VII Section 1.

The noise/error range can be estimated by $W = \max_{\theta_i \in \Theta}\left(\left|J(\theta_i) - \hat{J}(\theta_i)\right|\right)$ after adjusting for the mean values. We then select the neutral OPC class for this example. Once the good enough criterion g and the alignment level k are specified, the required selected subset size s from the crude model Eq. (2.44) is given by the function $Z(g,k/\text{neutral},W)$ in Table 2.1. Notice that these selected elements correspond to the first s members of N, because of the monotone property of the crude model. Then, we compare the selected subset with the true model to determine which indeed matches the good enough designs. 1000 experiments, each with a different N, are generated, so as to validate the actual observed alignment probability against AP = 0.95. We determined the alignment of each subset of size $s = Z(g,k/\text{neutral},W)$, where $g = 20, 30,\dots,200$ and $1 \leq k \leq 10$. Some of the alignment probabilities are plotted in Fig. 2.18. Each line in Fig. 2.18 represents the fraction of the 1000 experiments in which there are at least k of g good enough designs matched in the selected set. Note that:

1. The alignment probabilities are in general greater than 0.95, and this can be attributed to the conservative estimate of the function $Z(\bullet/\bullet)$.

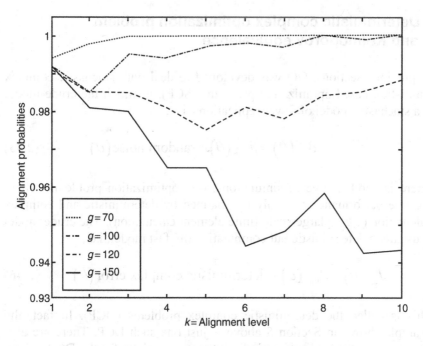

Fig. 2.18. Alignment probability validation for the example

2. Some fluctuation of the alignment probabilities are observed, which is due to the residues of the regression function Z.

The concept of universal alignment probability and the function $Z(\bullet/\bullet)$ have been validated many times in all papers on OO (Lau and Ho 1997; Shen and Bai 2005) and more examples will be shown later on in this book. Finally, we note that the blind pick AP of Eq. (2.37) of Section 5.1 is always a quick-and-dirty lower bound that is useful.

Exercise 2.12: Recall that in Section 2 we introduced 5 types of OPCs. Suppose we have a problem with many good designs (the flat type), and a problem with many bad designs (the steep type). Suppose the noise level is small. If we define the top-5% designs in both problems as the good enough designs, please use the UAP table just introduced to calculate the value of s such that $\text{Prob}[|G \cap S| \geq 1] \geq 0.95$. Which problem requires a larger selected set? Is this result counter intuitive? Shall we set the same value of g for both problems?

6 Deterministic complex optimization problem and Kolmogorov equivalence

In previous sections, OO was developed to deal with stochastic complex simulation-based optimization problems (SCP), in which the crude model is a stochastic model of fewer replications, i.e.,

$$J_{est}(\theta) = J_{true}(\theta) + \text{random noise}(\theta). \qquad (2.45)$$

There is another type of simulation-based optimization problems, where the true performance can only be obtained by deterministic and complex calculation (e.g., a large-scale finite element calculation). The crude model is usually a deterministic but computationally fast model, i.e.,

$$J_{est}(\theta) = J_{true}(\theta) + \text{deterministic complex error}(\theta). \qquad (2.46)$$

This is called the deterministic complex problems (DCP). In fact, the example shown in Section 5 above is just one such DCP. There are also many successful applications in both types, especially for the DCP, (Yang 1998; Guan et al. 2001; Lin et al. 2004) just to name a few. One question immediately arises:

In what sense are OO in DCP and OO in SCP equivalent s.t. the UAP table in Section 5 can be used in both cases?

We address this question in this section.

First, let us compare the two problem formulations in Eqs. (2.45) and (2.46). Digital computers have pervasive applications in simulation-based optimization. We cannot generate pure random numbers in a digital computer. Instead, we use pseudo random number generator (PRNG). When both the PRNG and the seed are fixed, all the numbers thus generated compose a deterministic and complex sequence. As long as either the PRNG or the seed is not known to the user, which is the case in any engineering practice, the number thus generated is unpredictable. Then tremendous amount of simulation literature (Fishman 1996; Yakowitz 1977; Gentle 1998; Landau and Binder 2000) have established that we can regard the number generated by a PRNG as a random number since they pass rigorous statistical tests. The concept of Kolmogorov complexity (Li and Vitányi 1997) also justified that we can regard the unpredictable deterministic number as a random number, which means that there is no fundamental difference between the two problem formulations in Eqs. (2.45) and (2.46), from an engineering viewpoint.[12]

Second, let us look at the application procedures for OO in SCP and OO in DCP (Box 2.2), which are *almost* identical.

There are three differences between the above two columns: step 2, 3, and 4. In Step 2 and 3, the differences are mainly about the names. The two Step 2's are equivalent in the sense that the performance evaluation is a complex and time-consuming calculation. The two step 3's are equivalent in the sense that a complex deterministic error and a random noise is equivalent w.r.t. Kolmogorov complexity, as aforementioned. We now focus on Step 4 and answer why the UAP table in Section 5 for SCP can be also used for DCP. Suppose we want to regress another UAP table for DCP. Then we need to repeat the experiments, exactly as we did in Section 5, when Θ is extremely large that almost no design can be selected more than once in the initial random sampling of N designs. Thus all the experimental data are statistically equivalent to those obtained when regressing the UAP table for SCP. So the table thus regressed should be

---

[12] In principle, for any DCP for which we wish to apply OO, we should go through the same rigorous statistical analysis as we have done in the simulation literature to establish that the errors can indeed be equated to random noises in Eq. (2.46). For engineering applications, we often take as an article of faith based on the Kolmogorov equivalence that the complex incompressible and unpredictable error sequence in Eq. (2.46) are indeed random. So far this assumption has worked in all references cited.

the same as the UAP table in Section 5, subject to statistic error. This is why we can use the same UAP table in both SCP and DCP.

Box 2.2. Comparison of the procedures for OO in SCP and for OO in DCP

OO in SCP	OO in DCP
Step 1: randomly sample N designs from Θ	Step 1: randomly sample N designs from Θ
Step 2: *stochastic crude model*-based performance evaluation	Step 2: *deterministic crude model*-based performance evaluation
Step 3: estimate OPC and the *noise level*. User specifies g and k.	Step 3: estimate OPC and the *error level*. User specifies g and k.
Step 4: calculate s using the UAP table	Step 4: calculate s using the UAP table
Step 5: select the observed top-s designs as S	Step 5: select the observed top-s designs as S
Step 6: The theory of OO ensures there are at least k good enough designs in S with high probability.	Step 6: The theory of OO ensures there are at least k good enough designs in S with high probability.

Readers may also consider the case when there are correlations among the deterministic errors in Eq. (2.46) for different designs. This can be regarded as correlated noise or independent non-identical noise, which will be addressed in Chapter VII, Section 3. Here we just summarize that it has already been shown by numerical experiments and theoretical explanations that the correlation among the noises seldom can hurt and actually helps most of the time (Deng et al. 1992). For the case of independent non-identical noise, there are ways to divide the designs into several groups, within each of which the noise are i.i.d. Then we can easily extend the method in this chapter to deal with the problem (Yang 1998).

In short, as long as J_{true} in Eq. (2.46) can be assumed to be Kolmogorov complex, we can apply OO to deal with the optimization problem

$$\min_{\theta \in \Theta} J_{\text{true}}(\theta), \tag{2.47}$$

given the crude model

$$J_{\text{est}}(\theta) = J_{\text{true}}(\theta) + \text{noise/error}(\theta). \tag{2.48}$$

We estimate the noise/error level, and the ordered performance curve. As long as the design space Θ is extremely large, we can use the UAP table (Table 2.1 in Section 5) to decide the appropriate selection size.

7 Example applications

7.1 Stochastic simulation models

Let us consider the cyclic server problem discussed in (Ho et al. 1992). The system has 10 buffers (of unlimited capacity) for 10 arrival streams modeled by Poisson processes with rates $\lambda_1, \ldots, \lambda_{10}$ respectively. There is a single cyclic server serving the 10 buffers in a round-robin fashion: at buffer i, m_i jobs are served (if there are less than m_i jobs in the buffer, then serve all the jobs until the buffer becomes empty); then, the server moves from buffer i to buffer $i + 1$ with a changeover time of length δ_i (Fig. 2.19). A holding cost of C_i units at buffer i is incurred. The objective is to find a service policy $(m_1, m_2, \ldots, m_{10})$ such that it minimizes the average holding cost per job per unit time in the system. We assume that $0 < m_i < 10$ for all i; in other words, no more than 10 jobs may be served at each buffer for any policy. The design space Θ is therefore the lattice

$$\Theta = \left\{ m = (m_1, m_2, \ldots, m_{10}) / 0 \leq m_i \leq 10, \forall i \right\}.$$

The cost coefficients and arrival rates are respectively

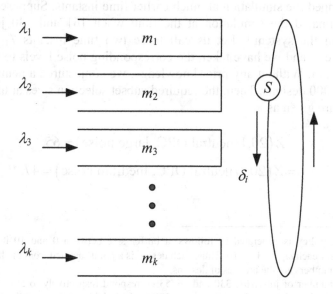

Fig. 2.19. Cyclic server serving K stream of arrivals

$$(C_1,...,C_{10}) = (1,1,1,10,1,50,1,1,1,1),$$
$$(\lambda_1,...,\lambda_{10}) = (1,1,1,1,1,1,1,1,1,1),$$

with a service rate of the server $\mu = 20$, and the mean changeover time of δ_i is

$$E(\delta_i) = 1/30, \text{ for all } i.$$

All random quantities are exponentially distributed. Notice that buffer 4 and buffer 6 have much higher cost coefficients.

We have generated 1000 policies (designs) from Θ and run long simulations for each policy to obtain their true ordering.[13] After 16753 jobs have arrived the system, the best 20 ordered designs are

$$\{\theta_{[1]},\theta_{[2]},...,\theta_{[20]}\} = \{761,166,843,785,417,456,205,925,234,70,586,91,93,$$
$$493,818,565,928,250,716,840\},[14]$$

which will be taken as the true ordering of the top 20 designs. Assume that we are interested in obtaining any of these top 20 designs; i.e., they form the good enough subset from the 1000 design samples; then, we could have stopped the simulation at much earlier time instants. Suppose that we had terminated the simulation at the time when 161 and 330 jobs had arrived in the system.[15] Let us call these two time instants T_1 and T_2, respectively, and we have taken the corresponding noise levels to be large and medium. Without any prior knowledge, we conjectured a neutral OPC for the 1,000 designs. Then, the required subset selection sizes at these two instants are given as

$$s_{T_1} = Z\left(20,1/\text{neutral OPC, large noise}\right) = 65,$$
$$s_{T_2} = Z\left(20,1/\text{neutral OPC, medium noise}\right) = 47.$$

---

[13] Each policy is generated as follows: a buffer size between 0 and 10 inclusive is generated for each m_i, $i = 1,...,$ 10. Thus, each design is a point sampled from the lattice Θ.

[14] The numbers are the indexes of designs.

[15] The number of jobs 161, 330 and 16753 correspond respectively to 500, 1000, and 50000 standard clock ticks. Simulation up to 50000 clock ticks is needed for the confidence intervals of the performance values of all designs to separate from each other. A standard clock tick is equivalent to an event happening to all 1000 systems operating under all policies. See Chapter VII Section b for further details about the standard clock.

Let us first examine the 65 designs at T_1,

S_{T_1} = {201, *166*, *565*, *818*, 702, 335, 487, 471, 73, 331, *843*, 172, 139, 595, 945, 905, 156, 658, 649, 431, 969, 233, 130, 204,307, 105, *840*, 29, 179, 189, 58, 305, 40, 38, 9, 525, 31, 286, 17, 366, 982, 914, 529, 655, 567, 828, 640, 621, 53, 301, 527, 924, 165, 459, 126, 597, 285, 643, *761*, 958, 681, 242, 379, 83, 927};

we see that six designs (in boldface and larger italics) are included in S_{T_1}. At T_2, the 47 selected designs are

S_{T_2} = {*761*, 595, *565*, 873, *843*, 139, 525, 105, *166*, *818*, 477, 643, 567, 447, 417, 980, 969, *234*, *928*, 366, 686, 201, 702, 738, 704, 111, 255, 314, 982, 361, *785*, 640, 773, 910, 901, 235, 455, *70*, 914, 172, 925, 335, 897, 31, *456*, 217, 176};

we see that ten designs (in boldface and larger italics) from the good enough subset have been captured. The true top-20 designs in order by definition are

{761,166,843,785,417,456,205,925,234,70,586,91,93,493,818,565, 928, 250,716,840}

It is also interesting to point out that, from our experiments, we have observed a very fast convergence of design orders. (See Section 4 for more details on the exponential convergence of ordinal comparison.)

7.2 Deterministic complex models

After the discussion in Section 6, we now can look at the example discussed in the end of Section 5 from another aspect. The optimization problem is to minimize

$$J(\theta) = a_1 \sin(2\pi\rho\theta) + a_2\theta + a_3, \qquad (2.49)$$

where $a_1 = 3$, $a_2 = 5$, $a_3 = 2$, $\rho = 500$, $\theta \in [0, 1]$. The deterministic crude model used to describe the increasing trend of $J(\theta)$ is

$$\hat{J}(\theta) = 5\theta. \qquad (2.50)$$

Since designs are taken randomly from the interval [0,1], the steps given in Section 5 are the steps to apply UAP table to decide selection set size s to solve this deterministic complex optimization problems. As seen from this example, by adopting a softened criterion, one can indeed achieve good alignment results by employing a very crude model in lieu of a complex model. This shows the importance of capturing the trend or general behavior of a system prior to the study of essential details. Perhaps, this also explains why a designer's intuition is often more valuable in the initial phase of a design process. Once a number of good enough designs are singled out, detailed studies of these designs can be done in the subsequent stages of the design process.

8 Preview of remaining chapters

So far, what we have presented in the first two chapters are introductory OO methodology and its foundations. Following the steps of OO and examples given, the readers can apply OO to solve real-world problems. In fact the majority of the 200 some references on OO employed no more than the theory and tools presented so far.

The remaining part of the book can be read more or less independently as shown in the logical dependency graph in Fig. 2.20 of the below. It is divided as three parts: Chapter III, IV, V and VI are major extensions of the OO method; Chapter VII deals with additional extensions; Chapter VIII presents case study for real-world examples.

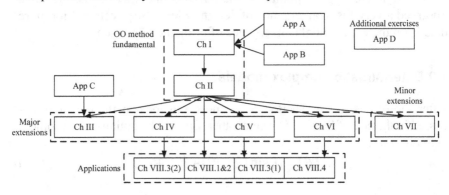

Fig. 2.20. Organization of the contents of the book

For the major extensions, we focus on Selection Rules in Chapter III. So far, we have studied two basic selection rules, namely, blind pick and horse race. We established analytical expression for blind pick and UAP table for the horse race. Although it is sufficient to use these rules to solve most application problems, it is still interesting to ask the natural question:

how about other selection rules? The purpose of Chapter III is to introduce more selection rules, compare the efficiencies of different selection rules, and give guideline in choosing selection rules based on the availability of computing budget.

As a second major extension to Ordinal Optimization method, we focus on optimization problems with multiple objective functions in Chapter IV. When there are multiple criteria (refers to as the vector case), ordinal comparison has to be done in a more complicated way than the scalar case of single objective function. As a result, the operative concept in multi-criterion optimization becomes the concept of Pareto optimum which was first formulated way back by Vilfredo Pareto. A design is said to be Pareto-optimal if it is not dominated by any other designs (i.e., there exists no other design that is better for at least one objective function value, and equal or superior with respect to the other objective functions). By introducing a natural order "layers" in design space, we generalize ordinal optimization from the scalar case to the vector case. We quantify how many observed layers are enough to contain the required number of designs in the Pareto frontier with high probability.

As a third major extension to Ordinal Optimization method, we focus in Chapter V on optimization problems with constraints. Similar to the objective function, we assume that the evaluation of constraints is also time consuming. So, the simple method of re-defining the design space as the feasible set then applying the tools of unconstrained OO does not work. To get around the time consuming evaluation barrier in constraints, we follow the idea of "crude model" in OO. Our key idea is to use a rough estimate of feasibility and allow the selected set to include some infeasible designs. Naturally to achieve the same level of alignment, more designs should be selected (thus a larger selected set is needed) for constrained OO. We quantify this additional correction.

A fourth extension to Ordinal Optimization method is given in Chapter VI. We deal with the memory limitation problem when we are trying to store a design on a computer. This problem comes naturally when we consider strategy optimization problems such as searching for good enough feedback control strategy for a complex system. Since for anything other than toy problems, the search space for admissible strategies can be enormously large, and the representation of a multi-dimensional function can be taxing on any size of computer memory, we need a way to search systematically in the strategy space that takes the limitation of memory storage into account. OO is incorporated into such a framework to search in the strategy space that can be implemented on a computer.

Further extensions of OO methodology requiring relatively little changes in solving real world problems will be discussed in Chapter VII.

Firstly, in previous study, no matter how large the design space is, we randomly sample $N = 1000$ designs and then apply OO to find some of the truly good enough designs (of these 1000 designs) with high probability. We show that the difference between the truly good enough designs (top-g%) of these 1000 designs, and the truly good enough designs (top-g%) of the entire design space is negligible for practical purpose. Thus further justify the practical use of OO for purists. Secondly, we show how we can take advantage of parallel computing ideas when applying OO to speed up the computation. The technique described is general. But the explanation is done by way of a specific example for clarity. Thirdly, in the previous consideration, only the i.i.d. additive noise is considered. However, the additive noise in practice might not be i.i.d. For example, the well adopted common random number generator technique is usually used in practice to reduce the variance of the observation noise. The observation noise then may be correlated. In some other times, the additive noise may be related to the performance of the solution, i.e., the noise is independent, but non-identical. We show that knowledge of these dependencies can help to improve the efficiency of OO method. Finally, as we mentioned earlier, Ordinal Optimization is not intended to replace the other optimization techniques. Instead, there are natural ways to combine ordinal optimization (including the key element of ordinal comparison) with other techniques, such as genetic algorithm to further improve the performance thus found.

In Chapter VIII, we present four real world application examples of applying OO method. The first example is a clothing manufacturing example. The problem is difficult and it is prohibitive to search for the best solution considering the tremendous computing budget involved. Using ordinal optimization ideas introduced in Chapter I and Chapter II, we obtained very encouraging results – not only have we achieved a high proportion of "good enough" designs but also tight profit margins compared with a pre-calculated upper bound. There is also a saving of at least 1/2000 of the computation time if brute-force simulations were otherwise used. The second real-world example is the Turbine blade design problem. We demonstrate how OO in Chapter I and II can be applied to solve such a deterministic complex problem. The third real-world example is the resource planning of a complex remanufacturing system involving two conflicting performance indices can only be evaluated by simulation. We demonstrate the application of Constrained Ordinal Optimization method developed in Chapter V and Vector Ordinal Optimization method developed in Chapter IV to the problem. At last, we demonstrate and apply extension of OO under limited memory developed in Chapter VI to the long standing strategy optimization problem known as the Witsenhausen Problem.

Chapter III Comparison of Selection Rules

In Chapter II, we learned that the selected set, S, plays an important role in the theory of OO. We also introduced two natural ways to determine the set S, namely, the Blind Pick (BP) and the Horse Race (HR) method. Of course there are also other ways possible. We have already mentioned in Section II.2 the example of how we take inspiration from the tennis tournament, where the pair wise elimination rule is used to determine the 16 surviving players (i.e., the set S) for the quarter final of the U.S. Open. Other rules for picking S come to our mind inspired by sports tournaments, such as the round-robin comparison imitating baseball (Jia et al. 2006a), which will be defined formally in Section 1. It is equally intuitive that different selection rules will lead to different Alignment Probability, Prob $[|G \cap S| \geq k]$, all other parameters remaining the same. In other words, to achieve the same high alignment probability, different selection rules may require us to select different number of designs for S. After we apply OO to find out the selected set S, which contains some good enough designs with high probability, we still need to use the detailed model to accurately evaluate the performance of the designs in S and finally choose the best one. It is clear that the larger the selected set S is, the more simulation is needed in the post selection comparison. So, it is of practical importance to find out which selection rule requires the smallest selection set in a given optimization problem. This selection rule will save the computational effort of the detail simulation the most efficiently. The purpose of this chapter is to make a careful study of the efficiency of different selection rules in terms of computing burden and efficacy in achieving high values of AP. For readers not interested in details and only concerned with the practical use of OO, s/he can go directly to the general recommendation at the end of this chapter which concludes that the HR is a generally good selection rule in the absence of detailed analysis and the BP rule can of course always serve as a lower bound in performance.

Since the size of the selected set is of the most practical importance, we use the size to evaluate and compare the performance of the selection rules. We will consider selection rule A better than B if selection rule A requires a smaller selected set S than B to achieve the same AP, all other parameters remaining the same. A natural question is: Is there a unique selection rule that is better than any others? If the answer is yes, then we only need to use this universal best selection rule. This is of course convenient for practical application of OO. Unfortunately, we have not found such a selection rule yet. Instead, each selection rule may have special advantages in dealing with certain problems. For example, as an extreme case, Blind Pick needs no problem information, nor any performance evaluation (not even a rough estimate based on the crude model). However, compared with Horse Race, BP usually requires a larger selected set to ensure the same high AP. So, a reasonable way to evaluate the performance of different selection rules is to classify different scenarios first, such as the problem types, noise levels, size of good enough set, required alignment level, and the computing budget we can use during the selection. Then we will be able to figure out what is the best selection rule in each scenario.

We have tried the above ways to compare different selection rules. However, after we obtain all the comparison results, i.e., which selection rule is the best in each scenario, there arises another question, that is, how can we present these results in an easy form? Obviously, the first idea might be using a lookup table. In other words, we list the scenarios and the corresponding selection rules that have the best performance within that scenario, and do this listing for all the scenarios. But there are too many scenarios. To get a rough estimate of how long this table will be, let us consider the following numerical example. We know there are 5 types of OPCs, and 3 noise levels (small, middle, and large). Suppose we consider 3 computing budgets (small, middle, and large), 20 different sizes of the good enough set (top-1%, 2%,...,20% designs), and 10 different alignment levels ($k = 1,2,...10$). Then the total number of scenario would be: $5 \times 3 \times 3 \times 20 \times 10 = 9000$. Suppose we have a table with 9000 lines, and the problem that we want to apply OO to solve fits in only one line, it would be inconvenient to find the line that is useful. We have to find better ways to present the comparison results.

We are fortunate to find this easy way to present the comparison results. The idea is to use a function with a few parameters to approximate the size of the selected set. Then giving the scenario, to which the practical problem belongs, and a selection rule, we use the function to approximate the size of the selected subset thus required. We do this for each selection rule. By comparing these approximated selection sizes, we then identify a predicted best selection rule in that scenario. As long as that prediction is good, i.e., either it is one of the several truly best selection rules in that scenario, or it is close to the truly best selection rule and the difference between the selected sizes is not large, then this predicted best selection rule is a truly good enough selection rule to be easily used in practice. We found such functions for many selection rules, including the ones that have already been frequently used in OO and some new ones. And we use two examples to show how this helps to find a good enough selection rule, and thus further save the computing budget.

The selection rules differ from each other considering how the computing budgets are allocated among the designs, how the designs are evaluated, and how the designs are finally selected. We should realize that there are a huge number of different selection rules. Most of these selection rules do not have a name yet, but they are different from the well-known rules such as BP and HR in the above sense. To get a rough idea, suppose there are all together $N = 1000$ designs, and we can evaluate these designs by at most $T = 30000$ simulations in total before selecting the subset. This equals to allocate these 30000 simulations to the 1000 designs. The number of all the allocation is

$$\begin{bmatrix} T+N-1 \\ N-1 \end{bmatrix} = \begin{bmatrix} 30999 \\ 999 \end{bmatrix} \gg 10^{1000},$$

which is a huge number. This example shows us two points. First, it is impractical to list and compare all the selection rules. Second, the selection rules we know is only a small portion of all the selection rule. So it is important to find some general properties, say selection rules with these properties are better (or have a large chance to be better) than others. These properties will help us to search for better selection rules in the future. We have found some these properties through theoretical analysis (Jia et al. 2006a; Jia 2006) and experimental study. One of them has justified that HR is in general a good selection rule.

The rest of this chapter is organized as follows. In Section 1, we classify the selection rules (especially the selection rules that are frequently used in OO literature) into different classes. Under mild assumptions, we establish a basic property of "good" selection rules. In Section 2, we use the selection size, s, to quantify the efficiency of the selection rule, and introduce a regression function to approximate this size for each commonly used selection rule. Through numerical experiments, we find that the comparison of the efficiency based on the regression function is almost exactly the same as the comparison result based on extensive simulation experiments. This justifies that the regression function is a reasonable tool to compare the efficiency of the selection rules. In order to show the readers how they can use the regression function to find a good selection rule and further reduce the selection size in practical problems, in Section 3, we consider two examples: a real function example and a queuing network example. In Section 4, we discuss three properties of good selection rules, which will help us to find better selection rules in the future. We summarize the application procedure in Section 5 and give some simple, quick and dirty rules to choose a good selection rule without detailed calculation and analysis.

1 Classification of selection rules

Let us now have a review of how a selection rule works. First, there is a crude model of the designs, which is computationally fast. In the extreme case of Blind Pick, no crude model is used. From an engineering viewpoint, this can also be regarded as using a crude model with infinite observation noise. So any observed order among the designs happens with equal probability. Second, based on the rough performance estimate thus obtained, a selection rule compares and orders the designs, and finally selects some designs as the selected set. Because the accuracy of the crude model used in the first step affects the size of the selected set dramatically, in the comparison of selection rules in this chapter, we assume the selection rules use the same crude model. Except in BP, a crude model with infinite observation noise is used. Note that different selection rules may use the rough performance estimate obtained through the same crude model in different ways. For example, a selection rule may compare designs in pairs, like in the U.S. tennis open, and regards all the designs in the quarter final as observed good enough designs. These 16 designs may be different from the observed top-16 designs, if selected by the Horse Race, because the observed second design may compete with the observed best design in an early round and fail, and thus cannot move into the quarter final. Also note that different designs may receive a different number of observations during the selection procedure. For example, in a season in National Basketball Association (NBA), each team plays the same number of games. If we regard the outcome of each game as an observation on the performances of the teams, each team receives the same number of observations. However, in the U.S. tennis open, if a player fails in an early round, he/she does not have the chance to move into the next round, which means finally the player may receive a different number of observations. Obviously, in simulation-based optimization, each observation of a design takes some part of the computing budget. That means different selection rules may assign the computing budget to the designs in different ways.

In summary, each selection rule in OO must answer two questions:

1. How to select S? – by ordering all N designs via either (a) or (b) below and select the top-s.

   (a) Using the estimated cardinal value no matter how crude it is.

(b) By the number of "wins" accumulated from the comparison of the estimated cardinal values no matter how crude they are. Two examples of the comparisons are:
(i) Comparison done pair-wisely
(ii) Comparison done globally.
2. How much computing budget is assigned to a design? – by either (a) or (b).
(a) Predetermined and allocated once.
(b) Successive iteration after initial assignment
(i) With elimination (i.e., some design will receive no more computing budget after certain iteration)
(ii) Without elimination.

Using answers to the above two questions, we consider and classify the following selection rules that are frequently used in ordinal optimization.

- **Blind pick (BP):** Assumes no knowledge of the problem and uniformly pick up s designs to make up S, i.e., (Question 1) random pick and (Question 2) no budget assigned (predetermined).
- **Horse race (HR):** The numbers of independent observations allocated to all designs are equal. By comparing the sample average, the observed top-s designs are selected. (Question 1) (a) and (Question 2) (a).
- **Round robin (RR):** Every design compares with every other design pair-wisely. In each comparison, we use only one observation (or equivalently "replication" in simulation language) per design to estimate the performance. Upon we completing the comparisons, every design wins some number of comparisons, including zero. We sort the designs by the number of wins in decreasing order[1]. The first-s designs are selected. For example, if there are four designs: θ_1, θ_2, θ_3, and θ_4. We need 3 rounds of comparisons:

$$\text{Round 1: } \theta_1 \text{ vs. } \theta_2, \theta_3 \text{ vs. } \theta_4.$$
$$\text{Round 2: } \theta_1 \text{ vs. } \theta_3, \theta_2 \text{ vs. } \theta_4.$$
$$\text{Round 3: } \theta_1 \text{ vs. } \theta_4, \theta_2 \text{ vs. } \theta_3.$$

Assume θ_1 and θ_3 win in round 1; θ_1 and θ_2 win in round 2; θ_1 and θ_2 win in round 3. Then the four designs win 3,2,1, and 0 comparisons,

---

[1] When a tie appears, the orders of designs within a tie are randomly decided. This assumption is also held for all the other selection rules mentioned in this book.

respectively. The observed best design is θ_1 and the worst is θ_4. (Question 1) (b)-(i) and (Question 2) (b)-(ii).

- **Sequential pair-wise elimination (SPE):** This is the rule used in tennis[2]. Designs are initially grouped into many pairs. The winners of these pairs are grouped into pairs again. This continues until a final winner appears. We show one example in Fig. 3.1. Assume θ_1 and θ_4 win in round 1; θ_1 wins in round 2. Then θ_1 is the observed best design. (Question 1) (b)-(i) and (Question 2) (b)-(i).

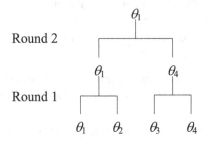

Fig. 3.1. One example of sequential pair-wise elimination (Jia et al. 2006a) © 2006 Elsevier

- **Optimal Computing Budget Allocation (OCBA)** (Chen et al. 2000): The idea is that we want to lavish larger computing budget on designs that more likely turn out to be good enough and not to waste efforts on designs that have a high likelihood of being bad. First, we randomly sample m_0 designs from Θ, and take n_0 observations of each design. Then we use a formula (Chen et al. 2000) to allocate additional Δ computing budget units to these m_0 designs to perform more replications. This procedure continues until all the computing budget is consumed. In OCBA we fix the "breadth"[3]. Section VII.4 has more details on this method. (Question 1) (a) and (Question 2) (b)-(ii).

- **Breadth vs. depth (B vs. D)** (Lin 2000b): The idea is to always allocate the next Δ computing budget units in the way that leads to a greater marginal benefit. There are two ways to get a marginal benefit: The breadth process, which means to sample new designs, and the depth

---

[2] In tennis tournament, the initial pairing is actually not totally random. We shall using total random pairing here since we assume no prior information on the designs.

[3] The "breadth" represents the number of designs explored in the optimization, and the "depth" represents the number of computing budget units allocated to each designs.

process[4], which means to do more observations of the designs that have already been sampled in order to get a better estimate of the design performance. The "benefit" of considering both breadth and depth is to increase the possibility of including truly good enough designs in the selected set. In B vs. D we can change the "breadth." (Question 1)(a) and (Question 2)(b)-(ii).

- **HR with global comparison (HR_gc):** In every round, the winners from the last round each receive one computing budget unit, and are compared with each other based on the new observations only. The observed best half designs win and the losers are eliminated in successive rounds. Finally we sort the designs by the number of rounds each design enters, from the largest to the smallest. (Question 1)(b)-(ii) and (Question 2)(b)-(i).

- **HR with no elimination (HR_ne):** In round i we compare the mean values of the observations so far, and allocate the additional Δ_i computing budget units to the observed best m_i designs. The value of Δ_i and m_i reduce by half each time. (Question 1)(a) and (Question 2)(b)-(ii).

- **HR as a counterpart of round robin (HR_CRR):** We allocate the computing budget as in RR. Finally we sort the designs by the average value of the observed data, not the number of wins. (Question 1)(a) and (Question 2)(b)-(ii).

We summarize the different rules in Table 3.1 below.

Table 3.1. Classification of selection rules

Selection Rule	How to select S?	Computing budget
BP	Random picking	No computing budget needed
HR	A	a
RR	b-i	b-ii
SPE	b-i	b-i
OCBA	A	b-ii
B vs. D	A	b-ii
HR_gc	b-ii	b-i
HR_ne	A	b-ii
HR_CRR	A	b-ii

Comparison and analysis in Section 2 and 3 will demonstrate that HR is in general a good selection rule, which is also a reason why we consider so

---

[4] For this rule we use BP in the breadth process and OCBA in the depth process. Other choices are possible and will yield different results (Lin 2000b).

many variants of the HR selection rules above (such as HR_ne, HR_gc, and HR_CRR).

Before going into details, we show here a basic property common to all comparisons of selection rules. Colloquially, we state it as

Seeing is believing – everything being equal (i.e., the same computing budget allocated to each design), we should always choose the observed top-s designs as the selected set.

First we introduce some notations. We use $\hat{J}(\theta_i)$ to denote one observation of design θ_i, i.e.,

$$\hat{J}(\theta_i) = J(\theta_i) + w_i = L(x(t;\theta_i,\xi)),   \qquad (3.1)$$

where w_i denotes the observation noise and $J(\theta_i)$ is the true performance value of design θ_i. We use $\overline{\hat{J}}(\theta_i)$ to denote the average observed value of the performance of design θ_i based on N_i independent observations (replications), i.e.,

$$\overline{\hat{J}}(\theta_i) = \frac{1}{N_i}\sum_{j=1}^{N_i} L(x(t;\theta_i,\xi_j))   \qquad (3.2)$$

where ξ_j is the randomness in the j-th sample replication. To simplify the discussion, we introduce the following assumptions:

Assumption 1: Observation noises of different designs are independently and identically distributed, i.e., w_i is i.i.d. for $i=1,2,...N$.
Assumption 2: We have no prior knowledge of the designs before conducting the simulations.

Now we state formally a basic property of the optimal selection rules.
Basic Property: Under Assumptions 1 and 2 and the equal allocation of computing budget to all designs, the selection of the observed top-s designs (by whatever rule) leads to an alignment probability no less than other selections on the average.

This property is intuitively reasonable and will be proved below. We give a short illustration in a specific case first to lend insight. Assume that there are only two designs θ_1 and θ_2 with true performances $J(\theta_1) < J(\theta_2)$, each design is observed only once, and the observation noises w_1 and w_2 contains i.i.d distribution. We define the truly better design θ_1 as the good enough design. And we can select only one design based on the observed performances. The alignment probability is simplified to the probability that we correctly select the better design θ_1. The question is: which design shall we select? In this case, the Basic Property says we should select the observed better one. To see why this is correct, we compare the probability that θ_1 is observed better $\text{Prob}\left[\bar{\bar{J}}(\theta_1) < \bar{\bar{J}}(\theta_2)\right]$ and the probability that θ_2 is observed better $\text{Prob}\left[\bar{\bar{J}}(\theta_1) > \bar{\bar{J}}(\theta_2)\right]$. For the first probability, we have

$$\text{Prob}\left[\bar{\bar{J}}(\theta_1) < \bar{\bar{J}}(\theta_2)\right] = \text{Prob}\left[J(\theta_1) + w_1 < J(\theta_2) + w_2\right]$$
$$= \text{Prob}\left[w_1 - w_2 < J(\theta_2) - J(\theta_1)\right].$$

Similarly, we can rewrite the second probability as

$$\text{Prob}\left[\bar{\bar{J}}(\theta_1) > \bar{\bar{J}}(\theta_2)\right] = \text{Prob}\left[w_1 - w_2 > J(\theta_2) - J(\theta_1)\right].$$

Since w_1 and w_2 are i.i.d., w_1-w_2 is with zero mean and has symmetric probability density function. Then due to the fact that $J(\theta_2)$-$J(\theta_1)$>0, it is obvious that

$$\text{Prob}\left[w_1 - w_2 < J(\theta_2) - J(\theta_1)\right] \geq \text{Prob}\left[w_1 - w_2 > J(\theta_2) - J(\theta_1)\right],$$

which means

$$\text{Prob}\left[\bar{\bar{J}}(\theta_1) < \bar{\bar{J}}(\theta_2)\right] \geq \text{Prob}\left[\bar{\bar{J}}(\theta_1) > \bar{\bar{J}}(\theta_2)\right], \qquad (3.3)$$

i.e., the truly better design has a larger chance to be better observed. If we follow the Basic Property, the alignment probability would be $\text{AP}_1 = \text{Prob}\left[\bar{\bar{J}}(\theta_1) < \bar{\bar{J}}(\theta_2)\right]$. If we do not follow the Basic Property, we must select the observe worse design with positive probability $q > 0$, then the alignment probability is

AP_2

$$= \text{Prob}\left[\text{select } \theta_1, \bar{\bar{J}}(\theta_1) < \bar{\bar{J}}(\theta_2)\right] + \text{Prob}\left[\text{select } \theta_1, \bar{\bar{J}}(\theta_1) > \bar{\bar{J}}(\theta_2)\right]$$

$$= \text{Prob}\left[\text{select } \theta_1 \middle| \bar{\bar{J}}(\theta_1) < \bar{\bar{J}}(\theta_2)\right] \text{Prob}\left[\bar{\bar{J}}(\theta_1) < \bar{\bar{J}}(\theta_2)\right]$$

$$\quad + \text{Prob}\left[\text{select } \theta_1 \middle| \bar{\bar{J}}(\theta_1) > \bar{\bar{J}}(\theta_2)\right] \text{Prob}\left[\bar{\bar{J}}(\theta_1) > \bar{\bar{J}}(\theta_2)\right]$$

$$= (1-q)\text{Prob}\left[\bar{\bar{J}}(\theta_1) < \bar{\bar{J}}(\theta_2)\right] + q\text{Prob}\left[\bar{\bar{J}}(\theta_1) > \bar{\bar{J}}(\theta_2)\right]$$

$$\leq (1-q)\text{Prob}\left[\bar{\bar{J}}(\theta_1) < \bar{\bar{J}}(\theta_2)\right] + q\text{Prob}\left[\bar{\bar{J}}(\theta_1) < \bar{\bar{J}}(\theta_2)\right]$$

$$= \text{Prob}\left[\bar{\bar{J}}(\theta_1) < \bar{\bar{J}}(\theta_2)\right]$$

$$= AP_1.$$

This means the Basic Property leads to an alignment probability no less than other selections. This also means "to see is to believe". When both designs are observed $N_1 = N_2$ times, since the noise of a single observation is i.i.d. for both designs, the difference between the average observed performance $\bar{\bar{J}}(\theta_i)$ and the true performance $J(\theta_i)$ (i.e., $\bar{\bar{J}}(\theta_i) - J(\theta_i)$) is also i.i.d. for both designs. Then we can extend the above proof to this case straightforwardly.

Exercise 3.1: Please extend the above proof to the case when both designs are observed more than once but equal times, i.e., $N_1 = N_2 > 1$.

We can use proof by contradiction to show the Basic Property in more general cases. First, note that when there are more than two designs, Eq. (3.3) still holds for any two designs (then the θ_1 in Eq. (3.3) represents the truly better one of the two designs). Suppose the selected set S, on the contrary, is not the observed top-s designs, then there must be some design θ that is within the observed top-s, but not selected; and there must be some other design θ' that is not within observed top-s, but is selected. Now we construct another selected set S'. The only difference between S and S' is: θ is in S' but θ' is not. Following Eq. (3.3), we know a truly better design θ has a bigger chance to be observed good enough. So we have

$$\text{Prob}\left[|S \cap G| \geq k\right] \leq \text{Prob}\left[|S' \cap G| \geq k\right]. \tag{3.4}$$

Following Eq. (3.4), each time we add to the selected set a design that is within the observed top-s and remove a design that is not, the AP does not decrease but only increase. If we keep on doing this, eventually we can obtain the selected set containing exact observed top-s designs whose alignment probability is no less than $\text{Prob}\left[\left|S \cap G\right| \geq k\right]$. This proves the Basic Property.

Exercise 3.2: when will the inequality in Eq. (3.4) be strict, and when not?

Astute reader might wonder: Is Assumption 1 critical to the Basic Property? The answer is yes. We show in the following an example, in which the observation noise is not i.i.d., and the Basic Property does not hold any more. Suppose there are three designs, θ_1, θ_2, and θ_3, with true performances $J(\theta_1)=0$, $J(\theta_2)=1$, and $J(\theta_3)=1.0001$. The observation noise is independent but contains non-identical normal distribution such that

$$\bar{\bar{J}}(\theta_1) \sim N\left(J(\theta_1), \sigma_1^2\right),$$

$$\bar{\bar{J}}(\theta_2) \sim N\left(J(\theta_2), \sigma_2^2\right), \text{ and}$$

$$\bar{\bar{J}}(\theta_3) \sim N\left(J(\theta_3), \sigma_3^2\right).$$

Let $\sigma_1 = 0.0001$, $\sigma_2 = \sigma_3 = 10$. Then we have

$$\text{Prob}\left[\theta_1 \text{ is observed as the best}\right]$$

$$=\text{Prob}\left[\bar{\bar{J}}(\theta_1) < \bar{\bar{J}}(\theta_2), \bar{\bar{J}}(\theta_1) < \bar{\bar{J}}(\theta_3)\right]$$

$$= \int_{-\infty}^{+\infty} \text{Prob}\left[\bar{\bar{J}}(\theta_2) \geq x\right] \text{Prob}\left[\bar{\bar{J}}(\theta_3) \geq x\right] p\left(\bar{\bar{J}}(\theta_1) = x\right) dx.$$

The last integration can be calculated using numerical integration. Similarly, we can calculate the probability that θ_2 or θ_3 is observed as the best. If we calculate the probability that each design is observed as the best, then we have

Prob[θ_1 is observed as the best]≈ 0.2914,
Prob[θ_2 is observed as the best]≈ 0.3543,
Prob[θ_3 is observed as the best]≈ 0.3543.

We can similarly calculate

Prob$[\theta_1$ is observed as the middle$]$

$$=\text{Prob}\left[\bar{\tilde{J}}(\theta_2)<\bar{\tilde{J}}(\theta_1)<\bar{\tilde{J}}(\theta_3)\right]+\text{Prob}\left[\bar{\tilde{J}}(\theta_3)<\bar{\tilde{J}}(\theta_1)<\bar{\tilde{J}}(\theta_2)\right]$$

$$=\int_{-\infty}^{+\infty}\text{Prob}\left[\bar{\tilde{J}}(\theta_2)<x\right]\text{Prob}\left[\bar{\tilde{J}}(\theta_3)>x\right]p\left(\bar{\tilde{J}}(\theta_1)=x\right)dx$$

$$+\int_{-\infty}^{+\infty}\text{Prob}\left[\bar{\tilde{J}}(\theta\ )>x\right]\text{Prob}\left[\bar{\tilde{J}}(\theta_3)<x\right]p\left(\bar{\tilde{J}}(\theta_1)=x\right)dx.$$

The numerical results are

Prob$[\theta_1$ is observed as the middle$]\approx 0.4968,$
Prob$[\theta_2$ is observed as the middle$]\approx 0.2516,$
Prob$[\theta_3$ is observed as the middle$]\approx 0.2516.$

In this case, if we follow the Basic Property and select the observed best design each time, then the alignment probability is 0.2914. However, if we select the observed middle design each time, then the alignment Probability is 0.4968. This means the Basic Property leads to a smaller alignment probability. This constitutes a counterexample if Assumption 1 is relaxed.[5]

Based on the Basic Property, in the following, our discussion will be based on Assumption 3 as we proceed below, if not mentioned explicitly:

Assumption 3. No matter what measurement and comparison method used, in every selection rule, the observed top-$|S|$ designs (under that specific measurement and comparison method) will be finally selected to compose set S.

We shall now compare selection rules in the following way:

---

[5]However, we should not take this counter example too seriously since the parameters used are most extreme. Almost all the analysis associated with OO in this book are very conservative in nature. Empirically in numerous applications, we and others have found that OO tools work well despite theoretical conditions being violated or impossible to verify. As we shall see later, in selection rules such as OCBA or B vs. D, assumption 1 is not satisfied. Noises are independent but not identical. Yet, in no cases, the analysis and prediction of this chapter did not work empirically.

Everything being equal (i.e., the same optimization problem, observation noise, computing budget, G, k, and α), which selection rule γ can use the smallest S to make $\text{Prob}\big[|G \cap S| \geq k/\gamma\big] \geq \alpha$?

2 Quantify the efficiency of selection rules

From the viewpoint of application, we may see the contribution of OO in this way: After screening (with or without iteration) using a crude model, a user is most interested in which of the subset of designs thus explored s/he should lavish her/his attention on in order to obtain a truly good enough design with sufficiently high probability. This selected subset denoted as S ideally should be as small as possible to minimize search and computing efforts. As we mentioned at the beginning of this chapter, we need to figure out what factors affect the size of the selected set, and by classifying all the scenarios, we compare the selection rules in each scenario and list the best for application. For a given high probability, α, the required minimum size for S is a function of the following parameters:

- $|G|=g$ – the size of the good enough set, e.g., the top-$n\%$
- k – the number of members of G guaranteed to be contained in S
- Noise/Error level, σ – in the estimation of the system performance using the crude model. We assume such noise is i.i.d. and can be roughly characterized as *small*, *medium*, and *large* with respect to the performance values being estimated. The case of correlated noise/error can also be handled (Deng et al. 1992), which will be discussed in Section VII.3.
- Problem Class, C – clearly if we are dealing with a problem contains many good enough solutions vs. a needle in a haystack type of problem, the required size of S can be very different. We express this notion in the form of an Ordered Performance Curve (OPC) which is by definition a monotonically nondecreasing curve that plots true performance value against all the designs ordered from the best to the worst (as defined in Chapter II and (Lau and Ho 1997; Ho et al. 1992; Ho 1999)). Five types of OPC are possible as illustrated in Fig. 3.2.
- Computing budget, T – the total number of simulations that can be used during the selection.

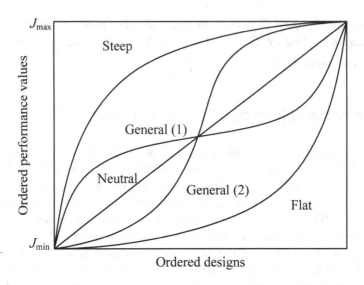

Fig. 3.2. Examples of ordered performance curves: flat, neutral, steep, and general

These factors can help distinguish a lot of different scenarios. As aforementioned, it is inconvenient for practitioners to use a very long lookup table to determine which selection rule is the best for a specific problem. So we need a function to approximate the size of the selected subset for each selection rule and in each scenario. In Section II.5, we introduced a regression function as follows:

$$Z(k, g) = e^{Z_1} k^{Z_2} g^{Z_3} + Z_4, \qquad (3.5)$$

where Z_1, Z_2, Z_3, Z_4 are constants depending on the OPC class, the noise level, g, and k values. When using the Horse Race selection rule and each design takes only one observation, Eq. (3.5) approximates the size of the selected subset pretty well. To test whether Eq. (3.5) can also be used to approximate the size of the selected subset for other selection rules and in other scenarios, we compare the true and the approximated values in each of the 9000 scenarios, for each selection rule. The detailed parameter settings of the experiments will be introduced in section 2.1. To get the true value of the size of the selected set, the idea is similar to that in Section II.5: we fix the true performance of each design, which then represents one of the five OPCs. Then we use a selection rule γ to determine which designs to select. Each time when the selection rule requires observing the performance of a design θ, we randomly generate the observation noise w and use $J(\theta)+w$ as an observed performance. If the selection rule assigns different numbers of replications to different designs, then the noises contained in observed performances will be different. All the available computing budgets are utilized in the way that is specified by the selection rule, and finally the selection rule obtains an order of all the designs according to the observed performances. Then we check whether there are at least k truly top-g designs contained in the observed top-s designs as selected by the selection

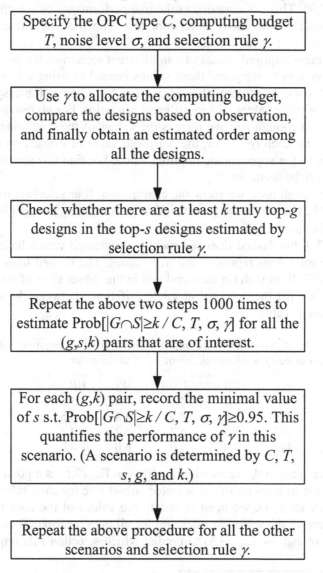

Fig. 3.3. The procedure to quantify the true performance of a selection rule

rule used. We check for each possible (g,s,k) values. Then we know whether a selection rule succeeds or fails in this experiment. We do 1000 experiments and then obtain the probability to succeed (as percent of the 1000 experiments) for each (g,s,k) values in that scenario. For each (g,k) pair, we record the minimal value of "s" s.t. the corresponding AP is no less than 0.95. This s is regarded as the true performance of selection rule γ in that scenario. The above procedure to obtain the true values of s in different scenarios is summarized in Fig. 3.3.

Giving these required values of s in different scenarios, we now want to find an easy way to represent these values instead of using a lookup table with 9000 rows. As aforementioned, we use Eq. (3.5) as the easy way to approximate these values of s in different scenarios. For all the (g,s,k) values with high enough AP, we regress the values of the coefficients in Eq. (3.5). Then for each (g,k) pair, the regressed value of s based on Eq. (3.5) is regarded as the approximate required value. We find this approximation is good in all the scenarios.

To get a rough idea, we show the examples of four selection rules (HR, OCBA, SPE, HR_ne) in the scenario of large computing budget (T=30000), neutral OPC, and middle noise level (σ = 0.5) in Fig. 3.4.

In Fig. 3.4, the dashed lines represent the regressed values based on Eq. (3.5). The solid lines represent the true values. The dashed lines approximate the solid lines well (in the trend and in the subset sizes of integer values of k). We list the maximal absolute difference between the solid and dashed lines in Table 3.2.

Table 3.2. The maximal absolute approximation error of required subset sizes (unit: number of designs) (Jia et al. 2006a) © 2006 Elsevier

g	HR	OCBA	SPE	HR_ne
10	3.8[6]	4.1	7.0	3.5
20	3.9	4.6	3.8	5.5
30	5.0	5.7	6.2	4.3
40	3.9	4.0	6.0	4.9

Since we see that the regression function in Eq. (3.5) is a good approximation to the true values of the selected subset size for each selection rule and in each scenario, we need to obtain the values of the coefficients in Eq. (3.5) for all the selection rules and for all the scenarios. Then giving a scenario, we can use Eq. (3.5) to predict which selection rule requires the

---

[6] The regressed subset size may not be integer, so the difference between the solid and dashed lines may not be integer. In application, we can just use the smallest integer that is no smaller than the regressed value as the subset size.

smallest subset to select. In Section 2.1, we introduce the parameter settings of all the scenarios in details. In Section 2.2, we discuss how we should compare selection rules using Eq. (3.5).

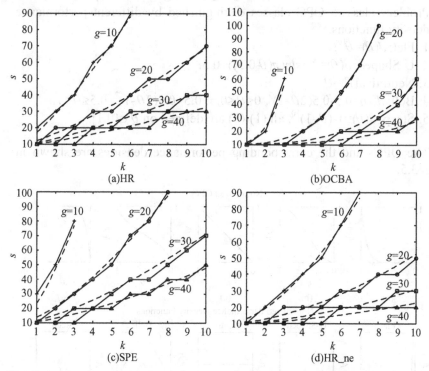

Fig. 3.4. The regressed values of Eq. (3.5) are good approximations of the true values of s. (Scenario: large computing budget, neutral OPC, and middle noise level) (subfigure (b) is from (Jia et al. 2006a) © 2006 Elsevier)

2.1 Parameter settings in experiments for regression functions

This subsection might be interesting only to readers who want to repeat our experiments. Other readers may want to skip this subsection and go directly to Section 2.2 and see how they can compare selection rules based on the regression functions.

In our experiments, we use the following parameter settings.

- $|\Theta|=N=1000$;
- $g \in \{10,20,\ldots 200\}$;
- $s \in \{10,20,\ldots 200\}$;
- $k \in \{1,2,\ldots 10\}$;

- noise level: Assume the observation noise is independently and identically distributed (i.i.d.) and has normal distribution $N(0,\sigma^2)$, where $\sigma \in \{0.25, 0.5, 1.5\}$ for small, middle, and large noises respectively;
- Problem class: 5 OPC classes distinguished by different performance density functions.
  1. Flat: $J(\theta)=\theta^{10}$.
  2. U-Shaped: $J(\theta)=0.5sin(\pi(\theta-0.5))+0.5$.
  3. Neutral: $J(\theta)=\theta$.
  4. Bell: $J(\theta)=0.5-0.5(2\theta-1)^2, 0\leq\theta\leq0.5; 0.5+0.5(2\theta-1)^2, 0.5\leq\theta\leq1$.
  5. Steep: $J(\theta)=1-(\theta-1)^{10}$, all (1)-(5) are defined on $\theta \in [0,1]$.

The OPCs and the corresponding performance densities are shown in Fig. 3.5.

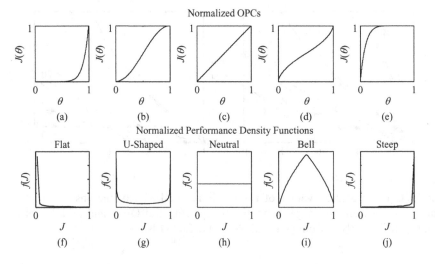

Fig. 3.5. Five types of OPCs and corresponding performance densities (Jia et al. 2006a) © 2006 Elsevier

- Computing budget: $T\in\{500,1000,30000\}$ for small, middle, and large computing budgets.
- "breadth" – the number of designs that can be explored by a selection rule. There is a close relationship between the "breadth" and the probability to find a good enough design. In one extreme case, a selection rule can maximize the "breadth" by observing each design only once. In this

way, the selection rule has a large chance to sample some truly good enough designs. However, since the observation noise is large, the selection rule may not be able to finally select these truly good enough designs. So a good selection rule should have a reasonable "breadth" and can adaptively change the "breadth" when the computing budget changes. The selection rules considered in this chapter require different computing budgets to explore the same number of designs (as shown in Table 3.3, where m_0 is the initial "breadth" and n_0 is the initial "depth" for each of these n_0 initial sampled designs. These formulas are obtained based on the definition of these selection rules in Section 1.). So when fixing the computing budget, the "breadth" of these selection rules are also different (as shown in Table 3.4).

- Other parameters: In OCBA and B vs. D, we try different values for n_0 and δ and find the following values works the best in the experiments, i.e., set $n_0 = 2$ and $\delta = 100$ when $T = 30000$; $n_0 = 1$ and $\delta = 10$ when $T = 1000$.
- For the combination of parameter settings above, we use 1000 independent experiments to estimate the alignment probability, which is expected to meet $\alpha \geq 0.95$.

Table 3.3. Computing budget units required by different selection rules to explore N designs (Jia et al. 2006a) © 2006 Elsevier

Selection rule	Computing budget units
BP	0
HR	N
OCBA	$\geq n_0 \times m_0$[7]
B vs. D	$\geq n_0 \times m_0$
SPE	$2N-2$
HR_gc	$2N-2$
HR_ne	$2N-2$
RR	N^2-N
HR_CRR	N^2-N

---

[7] m_0 and n_0 in OCBA and B vs. D are parameters controlled by the user. These two selection rules require $n_0 \times m_0$ computing budget units to roughly evaluate the m_0 randomly sampled designs at the beginning of the optimization, so the total required computing budget is at least $n_0 \times m_0$.

Table 3.4. The number of designs explored ("breadth") in different selection rules (Jia et al. 2006a) © 2006 Elsevier

Selection rule	Breadth		
	T=30000	T=1000	T=500
BP	1000	1000	500
HR	1000	1000	500
OCBA	1000^8	500	400
B vs. D	$\geq 10^9$	≥ 10	≥ 10
SPE	1000	501	251
HR_gc	1000	501	251
HR_ne	1000	501	251
RR	173	32	22
HR_CRR	173	32	22

Together with the blind pick selection rule, we tabulate all coefficients of the regression functions of different selection rules, as shown in Appendix C. It should be noted that since the subset sizes calculated by the coefficients above are used to approximate the experimental data, the values of g, s, and k should not exceed the above parameter settings of g, s, and k, which are used in the experiments, and the corresponding true AP Prob$[|G \cap S| \geq k]$ should be no smaller than 0.95. To be specific, the approximation has a working range of $20 \leq g \leq 200$, $s=Z(\bullet/\bullet)<180$, and the values of k and g should let the fraction k/g be small[10,11]. On the occasions where the noise factor is characterized to be within these predetermined levels (i.e., σ takes other values than 0.25, 0.5, and 1.0), proper interpolation

---

[8] Since the "breadth" of OCBA is fixed and set by the user, we tried different values. The values shown here are the best ones found in the experiments.

[9] B vs. D automatically changes the "breadth" based on the observed performance. So we only show the lower bound here, which is m_0 the initial number of designs sampled from the design space. The value of m_0 is set by the user. So we tried different values of m_0, and showed the best ones here.

[10] For B vs. D we have more constraints on the working range. When T=30000, the working range should meet $1 \leq k \leq 5$ and $k/g \leq 1/15$. When T=1000, the working range should meet $1 \leq k \leq 3$ and $k/g \leq 1/25$. When T=500, the working range should meet $1 \leq k \leq 2$ and $k/g \leq 1/35$.

[11] For RR and HR_CRR, we only list the regression values in the case of large computing budget (T=30000). In this case, RR and HR_CRR can explore 173 designs, and still have many choices of selected subset sizes. But when T=1000 and 500, the two selection rules can explore only 32 and 22 designs accordingly. To cover 1 or 2 of top-100 designs with probability no smaller than 0.95, we usually need to select all the explored designs, the number of which are 32 and 22, respectively.

of the subset sizes will suffice. When we need to predict the size of the selected subset in a scenario for a selection rule, we first find the table of that scenario and for that selection rule in Appendix C, then obtain the coefficients for the variables in Eq. (3.5). Then we use Eq. (3.5) to predict the value of *s*. Based on this prediction, we can compare different selection rules and find the best one.

2.2 Comparison of selection rules

Using these regressed values in the last subsection, we can easily predict the most efficient selection rule (among all selection rules of concern) in each scenario. As we will show by numerical examples in this subsection, the predicted best selection rule is usually one of the top-*n* selection rules. In the following, we consider one scenario as an example: large computing budget, middle noise level, and neutral OPC class. First, we use experiments to compare HR, OCBA, and B vs. D, the three selection rules that are frequently used in OO literature. As mentioned at the beginning of this section just prior to Fig. 3.3, we regard the size of the selected subset obtained by the experiments as true values. So we will regard the comparison result based on the experimental results as the true result. This is shown in Fig. 3.6 and explained below.

Fig. 3.6 shows in each (g, k) values, which selection rule is the best. Let us take a closer look at this figure, and discuss whether these comparison results are reasonable. First, the squares appear where the good enough set and the required alignment level *k* are small. When the good enough set is defined as the best design, the alignment probability $Prob[|G \cap S| \geq k]$ reduces to Prob[the observed best is the truly best], which is also known as the *probability of correct selection (PCS)* in OCBA. OCBA is developed to improve this probability. B vs. D is developed to improve Prob[the observed best design is truly good enough], which is close to PCS. Previous results in (Chen et al. 2000; Lin 2000b) also show that OCBA and B vs. D outperform HR in these cases.

Second, the black points appear where the good enough set is small, but *k* is large. To cover many good enough designs in the finally selected subset, we should make sure to explore a large number of good designs. Exploration, i.e., breadth, is thus more important than exploitation, i.e., depth. HR is developed to explore as many designs as possible, which is just the best choice in these cases. OCBA places emphasis on ensuring the observed best is the truly best, but does not pay attention to cover good enough designs. B vs. D takes both exploration and exploitation into consideration and tries to find a good balance. However B vs. D usually explores fewer designs

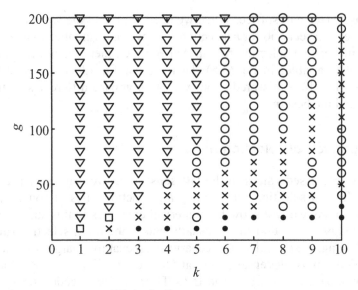

• HR is the best

× OCBA is the best

O HR and OCBA are better

□ OCBA and B vs. D are better

∇ HR, OCBA, and B vs. D are equal

Fig. 3.6. Comparison of HR, OCBA, and B vs. D under large computing budget, middle noise level, and neutral OPC class[12] (Jia et al. 2006a) © 2006 Elsevier

than HR, which does not make any additional effort in exploitation except for equally allocating computing budget units. So it is also reasonable that HR outperform OCBA and B vs. D in these cases.

Third, the crosses appear where the good enough set and the required alignment level are both of middle size/value. Both exploration and exploitation are important, and a good balance is needed. HR pays little attention to do the balance. It is difficult for B vs. D to outperform OCBA with a well-tuned "breadth" (Lin 2000b). This is why OCBA outperforms HR and B vs. D in many of these cases. Since the scenario in Fig. 3.6 has a large computing budget, even in HR each design obtains some exploitation (depth). So HR also performs best in some of these cases (represented by circles).

Fourth, the triangles appear where the good enough set is large and the required alignment level is small. These are easy cases and little effort in

---

[12] "the best" means this selection rule requires the smallest selected subset to achieve the same high AP in the same scenario. Again the experiment mentioned is the one associated with Fig. 3.3 and the text describing it.

balancing exploration and exploitation is needed. So HR, OCBA, and B vs. D require the same subset sizes in these cases.

Following the above analysis, we can summarize a couple of simple rules to choose a good selection rule among the three most frequently used selection rules in OO literature. For instance, in Fig. 3.6, a rule of thumb is: choose HR when the good enough set is small and the required alignment level is high; and choose OCBA in all other cases. These rules are also listed in the end of Section 5.

Now we predict the size of the selected set of these three selection rules using Eq. (3.5) together with the regressed value in Table C.20, C.21, and C.22 in Appendix C. By comparing these predicted values, we predict which selection rule is the best for each (g,k) values, as shown in Fig. 3.7.

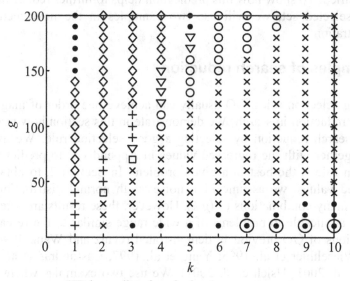

- ● HR is predicted as the best
- × OCBA is predicted as the best
- + B vs. D is predicted as the best
- ○ HR and OCBA are predicted as better
- ◇ HR and B vs. D are predicted as better
- □ OCBA and B vs. D are predicted as better
- ▽ HR, OCBA, and B vs. D are predicted as equal
- ⊙ Wrong prediction

Fig. 3.7. Comparison of HR, OCBA, and B vs. D using regressed values under large computing budget, middle noise level, and neutral OPC type (Jia et al. 2006a) © 2006 Elsevier

Note that in all (g,k) combinations, both Fig. 3.6 and Fig. 3.7 recommend a best selection rule. For a (g,k), if the recommended selection rule(s) in Fig. 3.7 is a subset of the ones recommended in Fig. 3.6, we say Fig. 3.7 gives a right recommendation; otherwise, we mark the corresponding case by a big circle. For example, in the bottom right of the figure, some points are marked. The reason is that the corresponding alignment probabilities are smaller than 0.95, as shown in Fig. 3.6. Note that in the aforementioned working range of $20 \leq g \leq 200$, $s=Z(\bullet/\bullet)<180$, and when the fraction k/g is small, Fig. 3.7 always gives **right** recommendations. This makes an easy way to choose a good selection rule once the computing budget to allocate (T), the noise level (σ), the OPC class, the size of good enough set (g), and the required alignment level (k) are specified. To show how this predication helps to further reduce the size of the selected set, we will use two examples in the next section as demonstration.

3 Examples of search reduction

By using selection rules, OO usually can achieve one order of magnitude or more reduction in search. We demonstrate in this section how to obtain further search reduction by selecting a good selection rule. We use Eq. (3.5) together with the regressed values in Appendix C to predict which selection rule is the best in a given problem. In Section 2, to obtain the regressed values, we assume i.i.d. noises with normal distribution, and explore many combinations of (g,k). However, these results are often universally applicable to problems of a wide range similar to the research in applications of OO previously demonstrated (Ganz and Wang 1994; Xie 1994; Wieselthier et al. 1995; Yang et al. 1997; Cassandras et al. 1998; Guan et al. 2001; Hsieh et al. 2001). We use two examples where either such assumptions are clearly violated or no knowledge concerning assumptions of the problem are available. In the first example, we test a function optimization problem, and show how to use an approximate model to achieve search reduction. In the second example, we test a practical manufacturing queuing network, and use the regressed values to compare HR, OCBA, and B vs. D.

3.1 Example: Picking with an approximate model

We consider again the example used in Section II.6, which shows how to apply our results when the accurate model is deterministic but complex.

We briefly review the problem formulation first. Consider a function defined on the range $\Theta = [0, 1]$

$$J(\theta) = a_1 \sin(2\pi\rho\theta) + a_2\theta + a_3, \qquad (3.6)$$

where $a_1 = 3$, $a_2 = 5$, and $a_3 = 2$. We set $\rho = 500$. So there are five hundred cycles in the range $[0, 1]$. The true model is deterministic, but rather complex considering relative and absolute minima. To get the exact model of function J and its minima, we need extensive samples of $\theta \in \Theta$, which we pretend to be costly. Instead, we find the trend for J is basically increasing, and then adopt a crude model

$$\hat{J}(\theta) = 50\theta. \qquad (3.7)$$

This is the linear part of function J. We regard the error between the true and crude models as noise, i.e.,

$$\hat{J}(\theta) = J(\theta) + \text{noise}. \qquad (3.8)$$

Assume that we have no prior knowledge on the noise, and that the noise has i.i.d. normal distribution $N(0, \sigma^2)$. By generating 1000 uniform samples of θ from $[0, 1]$

$$\Theta_N = \{\theta_1, \theta_2, \ldots, \theta_{1000}\},$$

we can estimate the standard deviation σ by

$$\sigma^2 = \text{var}_{\theta_i \in \Theta_N} \left(J(\theta_i) - \hat{J}(\theta_i) \right)$$

after adjusting for mean values.

We have totally 1000 computing budget units to allocate, i.e., we can take one observation per design in Θ_N on the average. HR will equally allocate the computing budgets to each design in this way. Other selection rules are different. In the following, we use the regressed values to estimate the minimal subset sizes of different selection rules. Note that the crude model in Eq. (3.7) is linear, so we choose neutral OPC class. After taking one observation of each design θ_i in, we find $\sigma = 0.5$ is a good estimate of the noise level. We try to cover at least k designs in the true top-50 ones in Θ_N, $k =1,2,\ldots 10$, with alignment probability no smaller than 0.95. Using Eq. (3.5) together with the regressed values in Table C.20 in Appendix C, we can estimate the minimal subset size of each selection rule, $\hat{s}$. Note that when evaluating designs, the selection rules use the crude model, and the true top-50 designs are defined by the complex model. To check whether $\hat{s}$ is a good approximate of the true value, we use experiments to test the true value, denoted as s. For a fixed value of s, we take 1000 independent samples of Θ_N to estimate the probability (the ratio of these 1000 experiments) that there are at least k truly top-50 designs in Θ_N contained in the observed top-s designs. Then we regard the minimal value of s to make AP≥0.95 as the true value. We compare the estimated value $\hat{s}$ with the true value s of each selection rule in Table 3.5.

Table 3.5. Estimated subset size of different selection rules in Example 3.1 (Jia et al. 2006a) © 2006 Elsevier

k	HR		OCBA		B vs. D		SPE		HR_gc		HR_ne	
	$\hat{s}$	s	$\hat{s}$	s	$\hat{s}$	s	$\hat{s}$	s	$\hat{s}$	s	$\hat{s}$	s
1	18	19	15	12	14	10	17	12	14	12	13	12
2	28	21	23	18	54	17	28	18	22	17	18	19
3	38	25	32	23	159	29	41	24	31	22	24	23
4	47	32	42	28			54	30	40	27	32	28
5	56	36	52	33			68	37	51	32	41	33
6	65	39	63	38			83	43	61	37	50	38
7	74	44	74	44			98	51	73	41	61	43
8	82	48	86	48			113	58	84	45	72	48
9	90	53	98	52			129	70	96	51	83	52
10	98	56	110	58			145	81	109	56	96	58

For B vs. D when we require to cover more than 4 top-50 designs, both the predicted subset size via the regression model and the true subset size exceed 200. So we do not list those corresponding sizes in Table 3.5. From Table 3.5 we can see that the estimated subset sizes are usually no smaller than the true sizes. This shows the conservative nature of the estimate. When the required alignment level k is specified, by comparing the estimated subset sizes, we can find the best among the selection rules. In Table 3.5, we find that HR_ne has the smallest estimated subset size among all the selection rules of interest for $k=1,2,\ldots 10$. So HR_ne is recommended in these cases. We can also sort the selection rules by the true subset sizes s in Table 3.5, from the smallest to the largest. We show the true order and the true AP of HR_ne among the selection rules for different k in Table 3.6.

Table 3.6. The true order and AP of HR_ne (Jia et al. 2006a) © 2006 Elsevier

k	True order	True AP
1	2	0.961
2	5	0.946
3	2	0.959
4	2	0.976
5	2	0.988
6	2	0.998
7	2	0.999
8	2	1.000
9	2	1.000
10	3	1.000

We compare 6 selection rules in this example. HR_ne is the predicted best one for $k = 1, 2,...,10$. From Table 3.6 we can see that this prediction is good, because HR_ne is within the top-2 selection rules in most k values, with the exception of $k = 2$ and 10. We analyze the two cases with more details. When $k = 2$, HR_ne is the 5th best selection rule. Please note, in that case, the true best subset size is 17, which is achieved by B vs. D and HR_gc. The subset size of HR_ne is 19, which is very close to 17. When $k=10$, HR_ne is the 3rd best selection rule, with a true subset size of 58. The true best subset size in this case is 56, which is achieved by HR and HR_gc. The two sizes are close to each other. Our method recommends HR_ne, which is a truly good selection rule in this case.

We can also test HR_ne from another aspect. Assume we take the recommendation, choose HR_ne as the selection rule, and use the regressed subset sizes accordingly for each value of k. We use 1000 experiments to estimate the true alignment probability (AP) that the selected subset of designs can cover at least k top-50 designs. The results are also shown in Table 3.6. From Table 3.6, we can see that the alignment probabilities are higher than the required value, 0.95, in most cases. There is only one exception, $k = 2$, in which AP is 0.946, very close to the required value 0.95. This example shows that the regression function can give conservative estimate of subset size and our method performs well in comparing selection rules. This is useful in practice. First we use the regressed value to estimate the subset sizes of different selection rules and find the best one. Using this best rule, we can obtain a subset of designs, which contains at least k true good enough designs with high probability. This often leads to a search reduction of at least one order of magnitude. As shown in this example, when the true model is complex, we can use a crude one in subset selections. Based on the crude model, using our method, we can recommend HR_ne as a good selection rule, which often is a truly good selection rule.

Exercise 3.3: Using the above method, we can easily find a good selection rule for a given problem. After we determine which selection rule to use, we can use the regressed values in Appendix C to calculate the size of the selected set for that selection rule. In this example, the estimated sizes of the selected set of HR_ne are shown in Table 3.5. However, when we take a close look at Table 3.5, we find the estimate $\hat{s}$ sometimes is too conservative from the true value s. How can we improve this estimate? [Hint: Section VIII.3 introduces a way to improve this estimate.]

3.2 Example: A buffer resource allocation problem

We consider a 10-node network as shown in Fig. 3.8. Such a network could be the model for a large number of real-world systems, such as a manufacturing system, a communication network, or a traffic network. For details about the background of this example, please refer to (Chen and Ho 1995; Patsis et al. 1997). We will test whether our method can recommend good selection rules on this practical model. There are two classes of customers with different inter-arrival distributions but having the same service requirement. We use c1 and c2 to denote the two classes of of customers. Their inter-arrival times are with uniform and exponential distributions respectively, i.e., c1: $U[2, 18]$, c2: $Exp(0.12)$. The 10-node network represents a connected three service stages. Node 0-3 denotes the first stage,

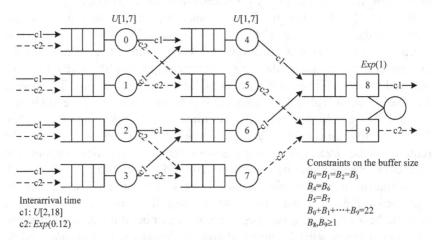

Fig. 3.8. 10-node network with priority and shared server (Jia et al. 2006a) © 2006 Elsevier

where customers enter the system. Node 4-7 denotes the second stage, where customers from different classes are served separately. Node 8-9 denotes the last stage, where customers leave the system after the service completes. The service time is with uniform distribution at node 0-7, i.e., $U[1,7]$, and is with exponential distribution with rate 1.0 at node 8 and 9, i.e., $Exp(1)$. All the nodes from 0 to 7 have their own servers, while node 8 and 9 share one server. Each node has a buffer with size B_i, $I = 0,1,...9$, (not including the one being served). It is the allocation of buffer size that makes this problem interesting. A buffer is said to be full if the queue length (not including the one being served) equals to the buffer size. If the buffer in the downstream is full when one node finishes serving one customer, this customer cannot leave and this server is idle and blocked. When two nodes are blocked by the same node in downstream, "first blocked, first served" is applied.

The two classes of customers have different priorities.

1. In queue of node 0-3, c1 customers jump before c2 customers. If a c1 customer arrives when a c2 customer is right in the process of being served, this service is allowed to be completed.
2. At node 8, when the queue length is greater than one, a c1 customer can be served. If a c2 from node 9 is in the process of being served while the queue length at node 8 becomes greater than one, the service to c2 will be interrupted to allow high priority (c1 customer).

For symmetric reason, we set the following constraints on buffer sizes:

$$B_0 = B_1 = B_2 = B_3$$
$$B_4 = B_6$$
$$B_5 = B_7$$
$$B_8 \geq 1, B_9 \geq 1.$$

We consider the problem of allocating 22 buffer units. There are totally 1001 different configurations. We want to find one configuration that can minimize the expected processing time for the first 100 customers, assuming there are no customers in the network initially.

For each of the 1001 configurations, we do 200 experiments and use the mean value to estimate the processing time for the first 100 customers. Although 200 experiments are not enough to discriminate the performance of different configurations exactly, the estimated top-100 designs are much more accurate than based on only one observation per design. We assume that there are totally 1001 computing budget units to allocate in the following experiments, as what we have mentioned above. So we sort the configurations using the estimated value and regard this as the true order.

We want to find at least k of top-g designs (e.g., $g = 100$, $k = 1,2,\ldots 10$) with high probability (e.g., $\alpha \geq 0.95$). Because HR, OCBA, and B vs. D are frequently used in ordinal optimization, we focus on comparing the three selection rules in this example. Since we have little knowledge of the performance of the network, we use neutral OPC class to approximate the true OPC of this system. From preliminary experiments, we find that 0.5 is a good estimate of noise level after adjusting for mean values. Assume that we have totally 1001 observations. Using the regressed values, we can estimate the subset sizes for each selection rule, $\hat{s}$. For each selection rule, we use 1000 experiments to estimate the true subset sizes, s, as we did in Section 2. We show the two groups of subset sizes in Table 3.7.

Table 3.7. Estimated and true subset sizes for Example 3.2 (Jia et al. 2006a) © 2006 Elsevier

k	HR		OCBA		B vs. D	
	$\hat{s}$	s	$\hat{s}$	s	$\hat{s}$	s
1	11	7	11	2	9	1
2	16	11	14	3	18	3
3	21	14	17	5	41	16
4	25	18	20	7		
5	30	22	24	10		
6	34	25	28	13		
7	38	28	32	16		
8	42	32	36	19		
9	46	35	40	23		
10	50	39	45	26		

When we use B vs. D to cover at least $k(k\geq 4)$ top-100 designs with a probability no smaller than 0.95, the true subset sizes(s) should be no greater than 200. So we do not list the values in Table 3.7. Recall also that the working range of the regression function in this case ($T = 1000$, middle noise level and B vs. D) is $1\leq k\leq 3$ and $k/g\leq 1/25$. So we only show the predicted subset sizes $\hat{s}$ and the true sizes s when $k = 1, 2$, and 3 for B vs. D. We also use 1000 replications to test the alignment probability for the estimated subset sizes of different selection rules, which we show in Table 3.8, regard as the true APs.

In Table 3.7, the estimated subset sizes $\hat{s}$ are usually greater than the true values s. In Table 3.8, the true alignment probabilities are always no smaller than the required value 0.95. This indicates that the estimated subset sizes are conservative and close to true values. We also list the predicted and truly best selection rule(s) under different k values, as is shown in Table 3.9.

Table 3.8. True alignment probability when using estimated subset size (Jia et al. 2006a) © 2006 Elsevier

k	HR	OCBA	B vs. D
1	0.994	1.000	0.998
2	0.999	1.000	0.993
3	0.999	1.000	0.975
4	0.999	1.000	
5	0.997	1.000	
6	0.999	0.999	
7	0.999	0.999	
8	0.999	1.000	
9	0.999	0.999	
10	0.999	1.000	

Table 3.9. True and predicted best selection rules for different k (Jia et al. 2006a) © 2006 Elsevier

k	Estimated best	Truly best
1	B vs. D	B vs. D
2	OCBA	OCBA and B vs. D
3	OCBA	OCBA
4	OCBA	OCBA
5	OCBA	OCBA
6	OCBA	OCBA
7	OCBA	OCBA
8	OCBA	OCBA
9	OCBA	OCBA
10	OCBA	OCBA

In Table 3.9, the predicted best selection rules are always a subset of the truly best ones. Thus, we can use the regressed value to predict what the best selection rule is. The recommended selection rule is truly the best.

Although our regressed values are obtained under the assumption of i.i.d. noise with normal distribution, it works well in other settings. For example, in Example 3.1 we regard the deterministic but complex error between the true model and the crude mode as observation noise. In Example 3.2, we do not know the true OPC type of the system and use neutral class as an approximate. From more experiments, we find the observation noise for different designs are not independently and identically distributed. However, numerical results have shown that our prediction works well (Jia et al. 2004).

Exercise 3.4: Are the symmetric constraints on the buffer size reasonable? If we remove the symmetric constraints on the buffer size, can we find a better solution?

4 Some properties of good selection rules

In Section 2 and 3, we have developed a method to predict the best selection rule among the nine considered in this chapter. However, as mentioned at the beginning of this chapter, there are a huge number of selection rules. The ones we discussed so far are only a small portion of all the possible selection rules. Understanding the method presented in the previous two sections can suffice the current practical application. It may also interest some readers if we can discover some general properties of good selection rules. These rules can then serve as a guideline for us to look for better selection rules in the future. In this section, we discuss three such properties, namely, *1) without elimination, 2) global comparison, and 3) using mean value of observations as the estimate of design performance.* (The three properties will be explained in details later in this section.) We use experiments to show that a selection rule with property 1) is no worse than others and actually is strictly better in 81% of all the tested scenarios; a selection rule with property 2) is no worse than others and actually is strictly better in 71% of all the tested scenarios; a selection rule with property 3) is no worse than others and actually is strictly better in 39% of all the tested scenarios. For those who want to repeat these numerical experiments, the detailed experimental data can be found in Section 4.4 in (Jia 2006).

Let's now consider the first property: without elimination. If a selection rule uses elimination, then a design that fails in an early round will have no chance to receive further observation, and will finally receive a low observed order. Sequential pair-wise elimination (SPE) is one of the selection rules that use elimination. HR_ne is not. To see how this property affects the performance of a selection rule, we compare SPE and HR_ne in all the five OPC types and three noise levels. Let there be a large computing budget. For $g = 20$, $s = 10, 20, \ldots, 200$ and different values of k, we use 1000 independent replications to test the AP in each case. The s value such that the corresponding AP is no less than 0.95 is the minimal size of the selected subset of that selection rule in that scenario. By comparing the sizes of the selected subset of SPE and HR_ne in all the scenarios mentioned above, we find HR_ne does not require a larger selected set than SPE in all the tested scenarios, and in 81% of the scenarios requires a smaller subset.

This justifies that without elimination is a property for good selection rules.

Then let's consider the second property: global comparison. A selection rule with global comparison will compare all the designs that enter a round together. A selection rule without global comparison will only compare some of the designs that enter a round. SPE uses pair-wise comparison, thus is an example of without global comparison. HR_gc compares all the designs in a round together, thus is a selection rule with global comparison. Since HR_gc and SPE both use elimination, by comparing these two rules, we can see how the global comparison affects the performance of a selection rule. Note that the Basic Property in Section 1 tells us that it does not hurt if we select the observed top-s designs. In each round HR_gc selects the observed top-half designs. This is consistent with the Basic Property. So, in principle, HR_gc should not be worse than SPE in all the scenarios. The question is: Since SPE is not consistent with the Basic Property, how much worse is this selection rule than HR_gc? We compare SPE and HR_gc in all the five OPC classes and three noise levels. Other parameter settings are also the same as when we used to compare SPE and HR_ne to test the first property. The result is: HR_gc is no worse than SPE in all the tested scenarios, and is better in 71% of the scenarios. This justifies that global comparison is a property of good selection rules.

Finally, we will consider the third property: using mean value of observations as the estimate of design performance. In some selection rules, the designs are ordered according to the number of "wins" they receive. RR, SPE, HR_gc are some of the examples. Some other selection rules order the designs according to the mean value of observations, e.g., HR and HR_CRR. To see how this property affects the performance of a selection rule, we compare HR_CRR and RR. Both selection rules equally assign the computing budget among the designs. Note, according to the Basic Property, selecting the observed top-s designs does not hurt in this case, which is exactly what HR_CRR does. So, in principle HR_CRR is no worse than RR. The question is: How much worse can RR be than HR_CRR? We compare HR_CRR and RR in all five OPC types and three noise levels. Let $g = 50$. There are large computing budgets. The comparison shows that HR_CRR is no worse than RR in all the tested scenarios, and is better in 39% of the scenarios. This justifies that using the mean value of observations as the estimate of design performance is a property of good selection rules.

5 Conclusion

We use regression functions to quantify the efficiency of different selection rules for ordinal optimization, especially HR, OCBA, and B vs. D, which are frequently used. Using the regressed values, we can predict the best selection rule(s) of interest under different parameter settings. The prediction is rather good, as we have shown in section 3 by numerical examples. We showed that some selection rules are no worse than some others, and discussed the properties of good selection rules from three aspects: Without elimination, global comparison, and using mean value as the measure.

To predict the best selection rule among a given set of selection rules using our method, we summarize the application procedure as follows (Box 3.1).

Box 3.1. How to predict a good selection rule

Step 1: Estimate the OPC class of the system and the noise level in observation.
Step 2: Specify the values of k, g, and α. We want to cover at least k designs in true top-g ones with high probability no smaller than α.
Step 3: Specify how much computing budget you have, i.e., you can simulate the system how many times during the entire optimization.
Step 4: Steps 1-3 determines a specific scenario. Find the regressed values of the coefficients in Eq. (3.5) in Appendix C for this scenario.
Step 5: Use Eq. (3.5) to predict the size of the selected subset for each selection rule. By comparing the predicted subset sizes, we can easily predict which selection rule is the best.
Step 6: Using the recommended selection rule, we can obtain a subset of designs, which contains at least k good enough designs with high probability.
Step 7: In this way, we can usually have a search reduction of at least one order of magnitude.

Based on ample experimental data, we also summarize a couple of simple, quick and dirty tips for easily picking up a good selection rule without calculation steps above. They are as follows:

1. In most of the cases, we recommend HR_ne, which works well and is a good selection rule among all the 9 selection rules compared in this paper.

2. In extremely difficult case (the computing budget is small, the size of good enough subset is small, and we try to cover many good enough designs), we recommend HR.

The above tips have intuitive illustrations. HR_ne has all the three properties of optimal selection rules mentioned in Section 4, which are without elimination, global comparison, and using mean value as the measure. Numerical results show each property can improve the alignment probability of selection rule separately. So HR_ne is a good choice in most cases. However, when the computing budget is extremely small and we can only explore part of the designs, exploration is more important than exploitation. Since HR usually explores a larger number of designs than HR_ne, it is a good choice in the latter difficult cases.

Since HR, OCBA, and B vs. D are the three frequently used rules in ordinal optimization, we also summarize some simple rules to choose a good one among them:

1. When the computing budget is small and we do not want much calculation to allocate the computing budget, HR is a good choice.
2. For special sizes of middle and large computing budget (i.e., there are 1000 designs and we have 1000 or 30000 computing budget units to allocate), OCBA is a good choice. Especially we prefer the following parameter settings in OCBA. When $T = 1000$, we prefer $m_0 = 500$. When $T = 30000$, $m_0 = 1000$.
3. In all other cases where we need a good and automatic balance between exploration and exploitation, B vs. D is a good choice.

Note that OCBA fixes the "breadth" and B vs. D can increase the "breadth" during the allocation procedure. So OCBA with an optimal "breadth" may have the same high alignment probability as B vs. D (Lin 2000b). This is why we prefer OCBA in the second rule above. We have used ample experiments to find the optimal "breadth" of OCBA in those cases. However, as pointed out in the third rule, in general cases we prefer B vs. D, which can do the balance automatically.

Finally and more generally, we can regard the selection rule as a way of narrowing down the search for a good enough solution. Currently, the practice of OO is first to narrow down from Θ to Θ_N. Denote this as the first stage. It typically uses uniform sampling to get a representative set from Θ. Then in the second stage we narrow down from Θ_N to S. This process is what we do with the selection rules, as has discussed in this chapter. However, this is only one possible way of doing the search for the "good enough". There still exist a number of other possibilities. For example, sampling in the first stage can be enhanced. Instead of uniform sampling, we can use heuristics to bias sampling towards more good designs. There need not be only two stages, but many stages leading to what may be called iterative OO (Deng and Ho 1997). In short, we can view each stage of the selection as overlaying a selection probability density over the set of design possibilities in question. Good selection rules are the ones that favor the good designs. More discussion on this will be found in Chapter VII.

Chapter IV Vector Ordinal Optimization

Consider a multi-objective optimization problem with m objective functions $J_1, \ldots, J_m$ over a finite but huge design space Θ. If the user knows the priority among these objective functions, or furthermore can assign appropriate weights to each objective functions, s/he can reformulate this problem as either a sequence of m single objective optimizations or a single objective optimization using the weighted sum of $J_1, \ldots, J_m$ as the objective function. Then the method introduced in Chapter II will suffice to solve this new problem. However, a more difficult case is that the user does not know the priority or the appropriate weights among the objective functions. In this chapter, we focus on this type of problem. The purpose of the optimization here is to find designs such that the objective functions are, in a sense, minimized. The operative concept in multi-criterion optimization problems, of course, is that of Pareto frontier or non-dominated solutions. All the designs in the Pareto frontier are considered Pareto optimal. The concept of Pareto optimum was formulated by Vilfredo Pareto (Pareto 1896). A design is said to be Pareto-optimal if it is not dominated by any other designs (i.e., there exists no other design that is better for at least one objective function value, and equal or superior with respect to the other objective functions). All Pareto-optimal points constitute the so-called Pareto frontier which plays the same role as maximum or minimum in single criterion optimization. As discussed before in Chapter I, exact values of the m objective functions are often computationally infeasible to obtain via simulation (due to the $1/(n)^{1/2}$ limit) and thus it is often hard to obtain the Pareto frontier. Genetic algorithms and evolutionary algorithms are alternatives (Goldberg 1989), which do not guarantee a set of designs in the Pareto frontier but try to find a set of designs hopefully not too far away from the Pareto frontier (Zitzler et al. 2003) and these methods do not consider the difficulty of time consuming simulation-based performance evaluation. Comprehensive surveys in this area can be found in (Coello 2000; Tan et al. 2002). In this chapter, by generalizing ordinal optimization from the scalar case to the vector case, we aim at quantifying how many observed layers (definition follows) are enough to contain the required number of designs in the Pareto frontier with high probability. Both tenets of the scalar OO are kept:

1. the order we introduced converges exponentially as the number of replications increases.
2. we ask for only some good enough designs that are pareto or nearly pareto optimal.

In Section 1, we first include a very brief review of the traditional vector optimization results for completeness. Then we define the concept of Pareto frontier and layer in vector optimization, the good enough set, selected set, universal alignment probability (UAP), and ordered performance curve (OPC) in the vector case. In Section 2, based on abundant experiments, we give the UAP table for a 2-dimensional case, quantifying subset selection sizes for different types of two-objective optimization problems. Following this idea, one can quantify subset selection sizes when there are an arbitrary number of objective functions. In Section 3, we show the exponential convergence rate of observed layers to true ones. In Section 4, we use examples to show how the above numerical results help to reduce the search efforts for true Pareto frontier by at least one order of magnitude. At last, we summarize in Box 4.1 the general steps to apply Vector Ordinal Optimization (VOO).

1 Definitions, terminologies, and concepts for VOO

First, we include here a very brief review of the traditional vector optimization problem of "Pareto-min $\sum_{i=1}^{m} J_i(\theta)$". To locate a point on the Pareto frontier, we usually resort to some form of scalarization of the vector criteria. The most popular method is to consider a weighted sum of the criteria

$$\min \sum_{i=1}^{m} \lambda_i J_i(\theta), \quad \text{with } \lambda_i \geq 0, \sum_{i=1}^{m} \lambda_i = 1, \qquad (4.1)$$

where the λ_i's play the role of LaGrange multipliers. Under appropriate convexity conditions, solutions of the scalarized problem over all λ_i can determine all points on the Pareto frontier. There are also other possible methods of scalairzation. For example, consider

$$\min \sum_{i=1}^{m} \lambda_i (J_i(\theta) - J_i^*(\theta))^2 \qquad (4.2)$$

where $J_i^*(\theta)$, called aspiration level, is the desired but unrealizable value of the ith performance criterion. Or consider,

$$\min\left\{\max J_i(\theta), i = 1, ..., m\right\}, \tag{4.3}$$

where we assume the true performances of any two designs are distinguishable in any objective function.

Exercise 4.1: Prove that solutions of Problems in Eqs. (4.1)-(4.3) above all result in a point on the Pareto Frontier.

However, our goals in VOO are somewhat different. We are not that interested in locating one point on the Pareto frontier. In this book, such a task is contrary to the basic tenets of OO. Instead, we want to locate a set of points which are "near or close to" the Pareto frontier as explained in the introduction part of this chapter.

Hence we now introduce definitions and notations necessary for VOO. These definitions parallel, in concept, to those defined in Chapter II for single objective ordinal optimization and can be understood similarly.

Θ the search space for the optimization variables θ.

J_i the performance criteria (also called objective functions) for the system. In contrast to the scalar case, we have m performance criteria, $i=1, .., m$.

N the number of designs uniformly chosen from Θ. It is understood that for each choice of θ, there corresponds a set of values $J_i(\theta)$, $i=1, 2, ... , m$.

$\prec$ the dominance relation between designs. A design θ_1 is said to dominate θ_2, denoted by $\theta_1 \prec \theta_2$, if $J_i(\theta_1) \leq J_i(\theta_2)$, for $i=1, 2, ... , m$, with at least one inequality being strict. If θ_1 does not dominate θ_2, θ_2 will be called noninferior to θ_1. Furthermore, if neither $\theta_1 \prec \theta_2$ nor $\theta_2 \prec \theta_1$ is true, θ_2 and θ_1 will be called incomparable.

$\mathcal{L}_1$ Pareto Frontier. A set of designs $\mathcal{L}_1$ is said to be in the Pareto frontier, in terms of the objective functions $J_1, ... , J_m$, if it contains all the designs that are not dominated by other designs in the design space Θ; i.e.,

$$\mathcal{L}_1 \equiv \{\theta | \theta \in \Theta, \nexists\, \theta' \in \Theta, \text{ s.t. } \theta' \prec \theta\}.$$

Designs in Pareto frontier are the counterparts of the true optimal design in the scale case.

Ω an operator that maps a design space to the set of the Pareto frontier with respect to the objective functions as $\mathcal{L}_1 = \Omega(\Theta)$. The concept of Pareto frontier can be extended to a sequence of layers.

$\mathcal{L}_i$ layers. A series of designs $\mathcal{L}_{i+1} = \Omega(\Theta \backslash \cup_{j=1,...,i} \mathcal{L}_j)$, $i=1, 2, ... $, are called layers, where $A \backslash B$ denotes a set containing all the designs included in set *A*, but not included in set *B*. Designs in $\mathcal{L}_i$ are called layer *i* designs. They are successive Pareto frontiers after the previous layers have been removed from consideration. *The significance of layers is that they introduce a natural order in the design space* Θ and there are *no preferences* on the objective functions and *no preferences* on the designs in the same layer.[1]

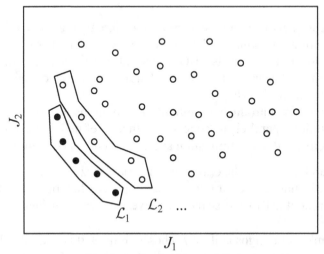

- ●   Designs in the Pareto frontier

- ○   Other designs

Fig. 4.1. Graphic illustration of layers (assuming minimization)

N_l the number of layers formed by the *N* designs uniformly chosen from Θ.

$\hat{N}_l$ the number of observed layers formed by the *N* designs uniformly chosen from Θ. This is a random number and varies in different replications.

$\hat{J}_i$ the observed performance criteria of the sampled designs. With *n* replications, we denote observed value of *i*-th performance criterion in

---

[1]There are other ways to introduce order for multi-objective optimization. For example, it was proposed in (Teng, Lee and Chew 2006) to sort designs according to the number of dominating designs. Pareto Frontier is the set of designs with 0 dominating designs. In that order, although there is no preference on the designs in the first layer (i.e., the Pareto Frontier), there usually are preferences on the designs in the second or other layers.

j-th replication by $\hat{J}_i(\theta, j) = J_i(\theta) + w_{ij}(\theta, \xi)$, $j=1,\ldots,n$, where w_{ij} are noises. By default, observed performance always refers to the average over all replications:

$$\bar{\hat{J}}_i(\theta) = \frac{1}{n}\sum_{j=1}^{n}\hat{J}_i(\theta, j), i=1, 2,\ldots, m.$$

Note, $\bar{\hat{J}}_i(\theta)$ is a random variable whose distribution also depends on the number of replications *n*.

$\hat{\prec}$ dominance in observation. A design θ_1 is said to dominate θ_2 in observation, denoted by $\theta_1 \hat{\prec} \theta_2$, if $\bar{\hat{J}}_i(\theta_1) \leq \bar{\hat{J}}_i(\theta_2)$, *i*=1, 2, . . . , *m*, with at least one inequality being strict.

$\hat{\mathcal{L}}_i$ observed layers. Dominance in observation is a stochastic relationship among designs, and will lead to stochastic partition of the design space into the observed layers $\hat{\mathcal{L}}_i$, *i*=1, 2,...

G   good enough set. Defined as the union of the designs in the true first *g* layers (e.g., when *g*=1,*G* is the Pareto frontier $\mathcal{L}_1$). As in scalar OO, the user is free to decide how many layers constitute *G*.

S   selected set. Defined as the designs chosen based on observed performances.

Selection Rule

Only a selection rule similar to the Horse Race rule is considered here. That is to select all designs in the observed first *s* layers.

G∩*S* the set of truly good enough designs in *S*.

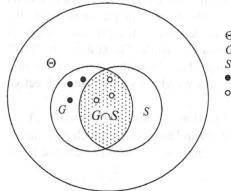

Θ : Search Space
G : Good Enough Set, the first *g* layers
S : Selected Set, the observed first *s* layers
● : Designs in Pareto Frontier
○ : Designs in observed Pareto Frontier

Fig. 4.2. Graphical illustration of Θ, *G*, and *S* in VOO

Alignment Probability (AP) $\equiv \text{Prob}[|G \cap S| \geq k]$

The probability that there are actually k truly good enough designs in S. This is the same as in scalar OO. Based on the notion of layers, it can be written as

$$AP = \text{Prob}\left[\left|\bigcup_{i=1}^{s} \hat{\mathcal{L}}_i \cap \bigcup_{i=1}^{g} \mathcal{L}_i\right| \geq k\right].$$

VOPC

Ordered Performance Curve in the vector case (VOPC). Similar to scalar OO, AP in VOO is also affected by problem types. We introduce VOPC for multi-objective optimization problems. VOPC is described by a function $F(x)$, where x is the layer index, from 1 to the total number of layers of that problem, and $F(x)$ is the number of designs in the first x layers. Correspondingly, we can also focus on the map $f(x)$, which sends the layer index x, ranging from 1 to the total number of layers of that problem, to $f(x)$, the number of designs in the x-th layer. In Fig. 4.3, we use two-objective optimization as an example to show how $f(x)$ describes different types of multi-objective optimization problems. There are three types of $F(x)$ in Fig. 4.3. Each column shows one type of two-objective optimization problem and the corresponding $f(x)$. The true performances of the designs are denoted by dots. Suppose that we uniformly pick up designs to compose the selected set S. In the first type, there are few designs in the Pareto frontier. Then, it is hard for S to contain some designs in the Pareto frontier. This type of optimization problems is hard. The problems in the second and third columns are neutral and easy, respectively. VOPC is a concept to classify the problem type, which is logically similar to OPC classifying the problem type in scalar OO. However, since we do not know the appropriate weight among the multiple objective functions, we cannot use the value of the objective functions to measure the "performance" of the designs in the same layer. Instead, we use the total number of designs in the previous layers as such a measure. Other definitions are possible. We still call it VOPC though we know the "performance" here is neither the value of any objective function nor the value of a weighted sum of these objective functions.

w_i noise/error level in objective function J_i. We assume $w_{ij}(\theta, \xi)$, $j = 1, 2, \ldots, n$, form an i.i.d. sequence of random variables with zero mean. When there is no confusion, we simply use w to represent the noise levels.

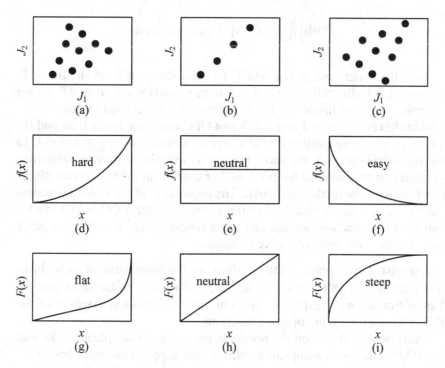

Fig. 4.3. Three types of two-objective optimization problems. A fourth type, the general type, is not shown

The Universal Alignment Probability (UAP)

$$\equiv \text{Prob}\left[|G \cap S| \geq k / N, N_l, w, \text{VOPC type}\right]$$

$$\equiv \text{UAP}\left(N, N_l, w, \text{VOPC type}\right)$$

As in scalar OO, the alignment probability can be tabulated once N, the number of designs, N_l, the number of layers, the noise/error level and VOPC type of a problem is given. The UAP table for the 2-dimension case will be given in Section 2.

Exercise 4.2: Please compare the concepts of order, good enough set, selected set, ordered performance curve, and universal alignment probability in ordinal optimization when there are one or multiple objective functions.

2 Universal alignment probability

In VOO, we care about the probability that the observed first s layers contain at least k designs of the true first g layers; we want this probability to be greater than or equal to some required confidence probability α, i.e.,

$$\text{Prob}\left[\left|\bigcup_{i=1}^{s}\hat{\mathcal{L}}_i\cap\bigcup_{i=1}^{g}\mathcal{L}_i\right|\geq k\right]\geq\alpha.$$

As in the scalar case, g, s, k, VOPC type and the noise level all affect AP. In general, it is difficult to get a closed-form formula to calculate AP, giving the values of these factors. In the scalar case, a table is used to quantify the relationship among g, s, k under different OPC and noise levels (Lau and Ho 1997). In VOO, we similarly tabulate the relationship among g, s, and k. In the rest of this section, we show how to do experiments on two-objective optimization problems as an example. For cases with more than two objective functions, the method is similar. The importance of the two-dimensional UAP table also lies in that, under mild assumption, the VOO-UAP table for two objective functions supplies an upper bound for the size of the selected set when there are more objective functions.

Exercise 4.3: Suppose all the designs are distinguishable in each objective function, i.e., $\forall\theta$, $\theta\in\Theta$, $i=1,2,...m$, $J_i(\theta)\neq J_i(\theta')$. Please show that the Pareto frontier with respect to m-1 objective functions is a subset of the Pareto frontier w.r.t. m objective functions.

Exercise 4.4: Based on the results in the last exercise, please show that the UAP table for two-objective optimization supplies an upper bound for the size of the selected set when there are more objective functions.

We consider three types of VOPCs in the experiments: Flat, neutral, and steep. Without loss of generality, we assume that the true performance of each design is within [0,1], that there are totally 10000 designs and 100 layers.[2] The numbers of designs in each layer are also specified:

---

[2]With no prior knowledge on the problem, for the neutral VOPC, we want to ensure the performance vectors of designs are uniformly deployed in an m-dimensional "cubic". So, fixing $N_l = 100$, for $m=2$, if there are 100 designs in each layer, there will be $N=100\times100$ designs, where N_l is the total number of layers. For the flat and steep VOPC, we want to ensure the performance vectors of the designs are uniformly deployed in an m-dimensional "pyramid". So, fixing $N_l = 100$, for $m=2$, let there be only one design in the first layer for the flat VOPC (or in the last layer for the steep VOPC), and the numbers of the designs in the successive layers increase (decrease in the steep VOPC) by a constant. For general m, to avoid the curse of dimensionality, fixing $N_l = 100$, we generate $N=C(1+N_l)N_l/2$ designs, and C is a positive number depending on m. See Fig. 4.4.

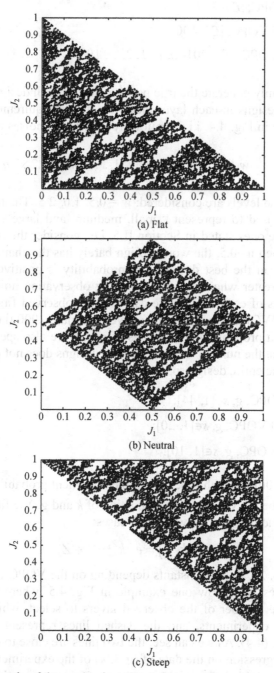

Fig. 4.4. One example of the randomly generated true performances of the designs for the three types of VOPC in the experiment

- for flat VOPC, $|\mathcal{L}_i|=2i-1$;

- for neutral VOPC, $|\mathcal{L}_i|=100$;

- for steep VOPC, $|\mathcal{L}_i|=201-2i$, $i=1, 2,..., 100$.

We randomly generate the true performance of each design such that the number of designs in each layer meets the above requirements. We show one example in Fig. 4.4. i.i.d. uniformly distributed noises are considered, i.e.,

$$w_{ij}(\theta, \xi) \sim U[-w,w], \; i=1, \ldots, m, j=1,2,...,n.$$

Three noise levels are considered: $w = 0.5, 1.0, 2.5$. The three noise levels are supposed to represent "small, medium, and large". The reason is similar to the ones stated in Section II.5, i.e., consider the neutral type for example, when $w=0.5$, the worst design barely has the chance to be observed better than the best design; this probability is positive when $w=1.0$, and much greater when $w=2.5$. By adding observation noises to the true performances of each design, we can find the observed first s layers. For each type of VOPC, we repeat the above procedure 1000 times to estimate the alignment probability. The values of g, s, k are also specified for each VOPC so that the number of good enough designs does not exceed 20% of the size of the entire design space:

- for flat VOPC, g, $s \in [1, 44]$;

- for neutral VOPC, g, $s \in [1, 20]$;

- for steep VOPC, g, $s \in [1, 10]$;

and $k \in [1, 100]$ for each type. When the alignment probability $\alpha \geq 0.95$, we try to describe the value of s as a function of k and g. We find that the following functional form fits well in all cases:

$$Z(k,g) = e^{Z_1} k^{Z_2} g^{Z_3} + Z_4, \tag{4.4}$$

where Z_1, Z_2, Z_3, Z_4 are constants depending on the VOPC types and noise characteristics. We show one example in Fig. 4.5, where the solid lines represent the number of the observed layers to select, which is obtained through the experiments, and the dashed lines represent the prediction given by Eq. (4.4). As we can see, the two lines are close to each other. We perform a regression on the data of (g, s, k) of the experiments, which lead to $\alpha \geq 0.95$, and this in turn produces the coefficients appearing in Eq. (4.4). We list the regressed values in Table 4.1.

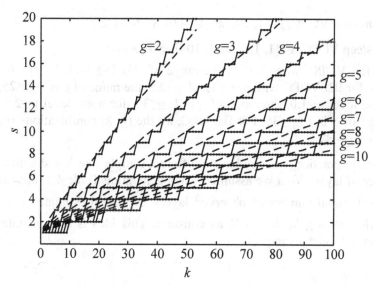

Fig. 4.5. The number of the observed layers to select to ensure AP≥0.95 in the Neutral VOPC and w=0.5. The solid lines represent the true value. The dashed lines represent the predicted value

Table 4.1. Regressed values of Z_1, Z_2, Z_3, Z_4 in $Z(k,g)$ for UAP of VOO

Noise	$U[-0.5, 0.5]$		
OPC class	Flat	Neutral	Steep
Z_1	4.2004	−0.2176	−0.7564
Z_2	1.1953	0.9430	0.9156
Z_3	−2.3590	−0.9208	−0.8748
Z_4	3.1992	1.0479	0.6250
Noise	$U[-1.0, 1.0]$		
OPC class	Flat	Neutral	Steep
Z_1	4.7281	0.3586	−0.1536
Z_2	1.0459	0.8896	0.8721
Z_3	−2.1283	−0.8972	−0.8618
Z_4	2.4815	0.8086	0.5191
Noise	$U[-2.5, 2.5]$		
OPC class	Flat	Neutral	Steep
Z_1	5.2099	0.9379	0.3885
Z_2	0.9220	0.8445	0.8536
Z_3	−1.9542	−0.8890	−0.8847
Z_4	1.9662	0.5946	0.5414

To ensure that Eq. (4.4) and Table 4.1 produces an upper bound estimate of the number of the selected layers, we restrict the numerical ranges as follows:

- for neutral VOPC, $g \in [1, 20]$, $k \in [1, 100]$, $s = Z(\cdot/\cdot) \leq 20$;

- for steep VOPC, $g \in [1, 10]$, $k \in [1, 100]$, $s = Z(\cdot/\cdot) \leq 10$;

- for flat VOPC, when $k \leq 50$, the range of g is $1 \leq g \leq 44$. When $50 < k \leq 100$, for flat VOPC with noise level $w = 0.5$, the range of g is $1 \leq g \leq 25$; for noise level $w = 1.0$, the range of g is $1 \leq g \leq 30$; for noise level $w = 2.5$, the range of g is $1 \leq g \leq 35$. For flat OPC, all the (g, k) combinations should let $s = Z(\cdot/\cdot) \leq 44$.

Note, in practice, we may have a different sample size N and a different number of layers N_l as we assumed when generating Table 4.1, our idea is to use the total number of observed layers $\hat{N}_l$ as an estimate of N_l and keep the ratios g/N_l, k/N, s/N_l as constant. This idea is demonstrated in Section 4 through examples.

3 Exponential convergence w.r.t order

VOO is based on ordinal comparison as in the scalar OO. That is, the comparison of designs and sorting them into observed layers. As in scalar OO, under mild conditions, it can be shown that ordinal comparison in VOO also has exponential convergence rate, namely the probability that the true i-th layer $\mathcal{L}_i$ is the same as the observed i-th layer $\hat{\mathcal{L}}_i$ is of form $1 - O(e^{-n\beta})$, where n is the number of replications and $\beta > 0$ is a constant. In fact, when some designs in the i-th layer $\mathcal{L}_i$ is not in the observed i-th layer $\hat{\mathcal{L}}_i$, there must be *at least one pair* of designs, the observed order (dominance) of which is different from the true order. Since the design space is finite, in order to prove the exponential convergence of layers, it is sufficient to show that any pair of designs θ_1 and θ_2 can only change their observed order with an exponentially decaying probability in terms of the number of replications n.

Prob[$\mathcal{L}_i = \hat{\mathcal{L}}_i$ for all i]

≥ 1-Prob[there exist a pair of designs θ_1 and θ_2 changing order in observation]

$\geq 1 - \sum_{\theta_1, \theta_2} \text{Prob} \left[\theta_1 \text{ and } \theta_2 \text{ changes order in observation} \right]$ (4.5)

Once an observation (based on n replications) is made, there are only three possible order relationships between any two designs θ_1 and θ_2:

$\theta_1 \prec \theta_2$, $\theta_1 \succ \theta_2$, and incomparable. For all these three cases, when a change in order happens in observation, there is at least one objective function J_i such that one of the following is true, among m objective functions.

1. $J_i(\theta_1) < J_i(\theta_2)$ and $\overline{\overline{J}}_i(\theta_1) \geq \overline{\overline{J}}_i(\theta_2)$ hold simultaneously.

2. $J_i(\theta_1) > J_i(\theta_2)$ and $\overline{\overline{J}}_i(\theta_1) \leq \overline{\overline{J}}_i(\theta_2)$ hold simultaneously.

In other words, in at least one objective function, a change in order occurs in observation. Thus, we can bound Prob[θ_1 and θ_2 changes order in observation] from above by Prob[$\overline{\overline{J}}_i(\theta_1) \geq \overline{\overline{J}}_i(\theta_2)$] when $J_i(\theta_1) < J_i(\theta_2)$ and Prob[$\overline{\overline{J}}_i(\theta_1) \leq \overline{\overline{J}}_i(\theta_2)$] when $J_i(\theta_1) > J_i(\theta_2)$.

It follows from the exponential convergence w.r.t. order for scalar case in Section II.4.2 (based on Large Deviation Theory) that probability for the order to change in one objective function decreases exponentially as a function of n. In other words, when $J_i(\theta_1) < J_i(\theta_2)$, and as long as the conditions on the samples (or equivalently on the noises) of scalar OO hold, i.e., the moment generating functions $E(e^{sw_{i1}(\theta,\xi)})$ exists for all $s \in (-d, d)$, for some $d > 0$, there must be a positive β such that

$$\text{Prob}\left[\overline{\overline{J}}_i(\theta_1) \geq \overline{\overline{J}}_i(\theta_2) \right] = O(e^{-n\beta})$$

and when $J_i(\theta_1) > J_i(\theta_2)$, there must be a positive β such that

$$\text{Prob}\left[\overline{\overline{J}}_i(\theta_1) \leq \overline{\overline{J}}_i(\theta_2) \right] = O(e^{-n\beta}).$$

As a result, we have

$$\text{Prob}\left[\theta_1 \text{ and } \theta_2 \text{ changes order in observation} \right] = O(e^{-n\beta})$$

and furthermore due to Eq. (4.5), we have

$$\text{Prob}[\mathcal{L}_i = \hat{\mathcal{L}}_i \text{ for all } i] = 1 - O(e^{-n\beta}).$$

Exercise 4.5: Another basic idea in single-objective OO is goal softening. What is the advantage to consider goal softening in VOO? Can we show some results similar to Section II.4.3?

4 Examples of search reduction

When we obtain the regressed values in Table 4.1, there are several assumptions:

(A1) There are 10000 designs and 100 layers in total.
(A2) The observation noises of different designs are independent.
(A3) The observation noises have uniform distribution.

It turns out that, even when these assumptions are not met, Table 4.1 still gives a good guidance on how many observed layers should be selected due to its universality. We present two examples here to demonstrate this. One is an academic problem, the other a practical problem. In Example 4.1, we relax assumptions A1 and A3 in Table 4.1. In Example 4.2, we relax all three assumptions.

4.1 Example: When the observation noise contains normal distribution

Consider a two-objective optimization problem $\min_{\theta \in \Theta} J(\theta)$, where $J(\theta) = [J_1(\theta), J_2(\theta)]^\tau$, $J_1(\theta)$ and $J_2(\theta)$ are the true performance of the design θ, and τ denotes transposition. There are 1000 designs in Θ. For each design θ, $J_1(\theta)$ and $J_2(\theta)$ are uniformly generated values from $[0,1]$ and are fixed in the following experiments. The true performances are shown in Fig. 4.6.

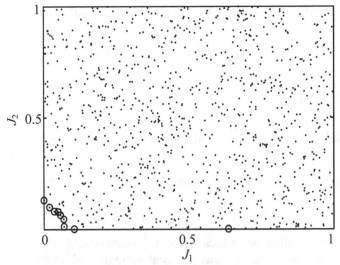

Fig. 4.6. True performances J_1 and J_2 in Example 4.1

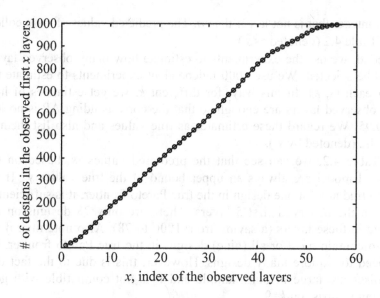

Fig. 4.7. The observed VOPC of Example 4.1

There are 9 designs in the Pareto frontier, which are marked by circles. The observation noise of each design is independent and has normal distribution $N(0, 0.252)$. We are interested to find at least k, $1 \leq k \leq 9$, designs in the true Pareto frontier with high probability, $\alpha \geq 0.95$. The question is how many observed layers we should select. In the following, two methods are compared.

First, we use the regressed values in Table 4.1 to answer this question. We simulate each design only once and estimate the VOPC type of this problem, which is neutral (as shown in Fig. 4.7). We specify the noise level as 0.5. Then, from Table 4.1, we find the values of coefficients as

$$Z_1=0.2176, Z_2=0.9430, Z_3=0.9208, Z_4=1.0479.$$

Since there are only 1000 designs and 579 observed layers in Example 4.1, we need to adjust the values of g and k. We keep the ratios g/N_l, k/N, s/N_l as constant, where N_l is the total number of true layers. We use the total number of observed layers $\hat{N}_l$ as an estimate of N_l. Then, we have

$$g' = \lfloor (100/57) \times 1 \rfloor = 1,$$
$$k' = (10000/1000) \times k = 10k, \ 1 \leq k \leq 9,$$

where $\lfloor a \rfloor$ represents the smallest integer that is not smaller than a. Using Eq. (1), we get $s'(k', g')$ and $s = \lceil (57/100) \times s \rceil$, where $\lceil a \rceil$ represents the largest integer that is not larger than a. The predicted values of s are collected in Table 4.2 (denoted as $\hat{s}$).

Second, we use the experiments to estimate how many observed layers should be selected. We use 1000 independent experiments to estimate the AP of each (s, k). In this way, for different k, we get estimates of how many observed layers are enough so that the corresponding AP is no less than 0.95. We regard these estimates as true values and also list them in Table 4.2 (denoted by s^*).

In Table 4.2, we can see that the predicted values $\hat{s}$ based on the regressed model are always an upper bound of the true values s^*. If we want to find at least one design in the true Pareto frontier, it is sufficient to focus on the observed first 5 layers. There are only 78 designs on the average in these layers (a saving from 1000 to 78). Also note that, if we want to contain most or all (nine) designs in the true Pareto frontier, we still need to explore many designs. However, this is due to the fact that the noises are large in our example and it is not compatible with goal softening to insist on $k=9$.

Table 4.2. Predicted and true values of s for Example 4.1

k	s^*	$\hat{s}$	$\left| \bigcup_{i=1}^{\hat{s}} \mathcal{L}_i \right|$
1	3	5	78
2	5	9	166
3	7	12	241
4	9	16	340
5	11	19	420
6	14	23	517
7	17	26	596
8	22	30	692
9	32	33	756

4.2 Example: The buffer allocation problem

We will consider a 10-node queuing network, as shown in Fig. 4.8. In fact, this example has already been introduced in Section III.3, but we considered only one objective function then. Now we are going to consider two objective functions (introduction follows) here. Let us briefly review the problem formulation. There are two classes of customers with different arrival distributions (exponential and uniform distributions). Both classes arrive at any of the 0–3 nodes and leave the network after finishing all three stages of services. The routing is class dependent and is deterministic.

The buffer size at each node is finite and is the parameter that we can design. We say that a buffer is full if there are as many customers as that buffer size, not including the customer being served. Nodes 8-9 have individual queues but share one server. This network can model a large number of real-world systems, such as manufacturing, communication, and traffic network. We consider the problem of allocating 22 buffer units among the 10 nodes. We use B_i to denote the buffer size at node i, $B_i \geq 0$. For symmetry reasons, we require

$$B_0=B_1=B_2=B_3, B_4=B_6, B_5=B_7, B_8, B_9>0. \qquad (4.6)$$

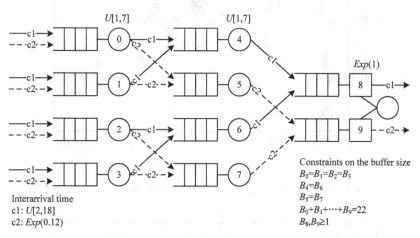

Fig. 4.8. The 10-node network with priority and shared server

We can get 1001 different configurations in all. There are two objective functions. One is the expected time to process the first 100 jobs from the same initial state (all buffers are empty). The other is the average utility of the buffers at all the nodes, i.e., $\sum_{i=0}^{9} q_i / B_i$, where q_i is the expected queue length at node i, $0 \leq i \leq 9$, where for $B_i = 0$, we define the utility of that buffer to be 1. We want to improve the throughput of the network and improve the efficiency of all the buffers. We formulate the problem as a two-objective minimization problem, where J_1 is the first objective function above and $J_2 = 1 - \sum_{i=0}^{9} q_i / B_i$.

For each design (a configuration of buffers) θ, we use 1000 independent experiments to estimate $J_1(\theta)$ and $J_2(\theta)$. The experimental results are shown in Fig. 4.9. We regard these values as true performances and define the configurations in the observed first two layers as good enough (9 designs in total), also marked by circles in Fig. 4.9. We want to find at least k, $1 \leq k \leq 9$, configurations in the true first two layers with high probability, $\alpha \geq 0.95$. The question is also how many observed layers should be selected.

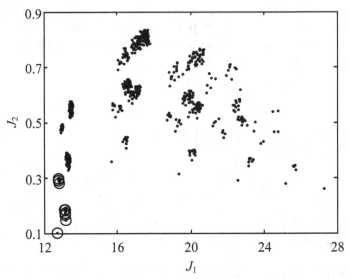

Fig. 4.9. The true performance of the configurations in Example 4.2

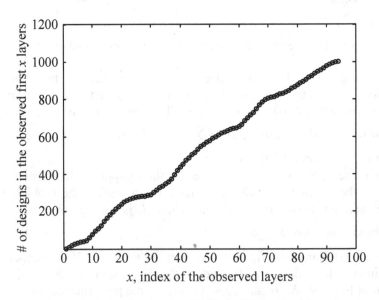

Fig. 4.10. The observed VOPC of Example 4.2

First, we simulate each configuration once (i.e., one replication only – a very crude estimate of the performance of the particular configuration). We show one instance in Fig. 4.10. There are 94 observed layers, which may be different in various experiments. The estimated VOPC type is neutral. By normalization, the standard deviation of the observation noise is 0.1017 for J_1 and 0.0271 for J_2, and we choose $0.5 > 2 \times 0.1017$ as the noise level. The corresponding coefficients in Table 4.1 are the same as those in Example 4.1. We adjust the values of g and k accordingly, i.e.,

$$g' = \lfloor 100/94 \times 2 \rfloor = 2, \ k' = (10000/1001) \times k \approx 10k, \ 1 \leq k \leq 9.$$

Substituting these values into Eq. (4.4), we can get $s' = Z(k', g')$ and $s = \lceil 94/100 \times s' \rceil$. We show the predicted number of observed layers to select in Table 4.3.

Second, we use 1000 independent experiments to estimate the AP of each (s, k). For each k, when AP is no smaller than 0.95, we denote the value of s as s^* in Table 4.3.

Table 4.3. Predicted and true values of s for Example 4.2

k	s^*	$\hat{s}$	$\left\| \bigcup_{i=1}^{\hat{s}} \mathcal{L}_i \right\|$
1	1	5	32
2	2	8	44
3	2	11	95
4	3	14	147
5	4	17	197
6	5	20	240
7	6	23	268
8	8	26	279
9	9	29	286

If we want to contain at least 3 designs in the true first two layers, according to Table 4.3, we need to explore only 95 designs on average. This saves much of our search efforts. We can see that the predicted values of s are always no less than the estimated values. The predicted values of $\hat{s}$ seem conservative. The reason is that the normalized noise level is 0.2034 in Example 4.2, which means that some good enough designs almost always dominate some other designs in observation. However, the smallest noise level in Table 4.1 is 0.5, which is more than twice as large as the normalized. In turn, this leads to a conservative estimate of s.

Example 4.2 violates all the three assumptions: There are only 1001 designs (configurations) in total; the observation noise is not i.i.d., and does not contain uniform distribution. However, numerical results show that the regression function (0.4) and the regressed coefficients in Table 4.1 still give good guidance on how many observed layers should be selected. When there are more objective functions, we can design similar experiments to obtain the regressed values of the coefficients in Eq. (4.4). In practical applications, we can use more problem information to obtain better prediction on the number of observed layers to select (i.e., a tighter upper bound of the true values). This will be shown in Chapter VIII Section 3.

Exercise 4.6: In Example 4.2, we introduced constraints in Eq. (4.6) for symmetry reason. What if we relax these constraints? Can we find designs with better performances in both objective functions? Please try to apply VOO to solve Example 4.2 without the constraints in Eq. (4.6).

Exercise 4.7: Please think, before reading Section VIII.3, about effective ways to obtain less conservative estimate of the number of observed layers to select, when we have the observed performance of the randomly sampled N designs.

Box 4.1. The application procedure of VOO

Step 1:	Uniformly and randomly sample N designs from Θ.
Step 2:	Use a crude and computationally fast model to estimate the m performance criteria of these N designs.
Step 3:	Estimate the VOPC class and noise level. The user specifies the size of good enough set, g, and the required alignment level, k.
Step 4:	Use Table 4.1 (if m=2, or similar tables generated beforehand for general m) to calculate $s=Z(g, k$/VOPC class, noise level).
Step 5:	Select the observed first s layers as the selected set.
Step 6:	The theory of VOO ensures that S contains at least k truly good enough designs with probability no less than 0.95.

Chapter V Constrained Ordinal Optimization

We discussed single-objective optimization in Chapter II and III, and dealt with multiple-objective optimization in Chapter IV. All these belong to unconstrained optimizations. Since we usually meet constraints in practice, a natural question is how we could apply ordinal optimization in constrained optimization problems. Traditionally, optimization problems involving constraints are treated via the use of LaGrange multipliers (Bryson and Ho 1969). See also Eq. (4.1) in the introduction of Chapter IV. The duality between constrained optimization with vector optimization is best illustrated via the following diagram (Fig. 5.1).

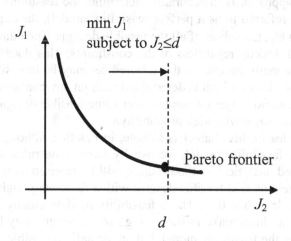

Fig. 5.1. The duality between constrained optimization and vector optimization

Setting different values of the parameter "d", we can determine various points on the Pareto frontier by solving a series of constrained optimization problem. Conversely, every point on the Pareto frontier solves a constrained optimization problem for some constraint value "d". Thus in principle, VOO and COO are also duals of each other. Chapter IV can be considered as a dual of this chapter.

More practically, in some cases, the constraints can be easily checked, e.g., simple linear inequality equations involving one or two variables. By modifying the design space, Θ, to one including only the feasible designs, Θ_f, we convert the original problem to an unconstrained one, which can then be dealt with by the methods in Chapter II-IV. However, in some other cases, it is time-consuming to check the constraints. In this chapter, we focus on the optimization problem of form

$$\min_{\theta \in \Theta} J(\theta) \equiv E\left[L(x,\theta,\xi)\right]$$
$$\text{s.t.} \quad h_i(\theta) \equiv E\left[L_i(x,\theta,\xi)\right] \leq 0, i = 1,...,m. \tag{5.1}$$

Considering the problems handled by OO in previous chapters, where the evaluation of $J(\bullet)$ is time consuming, we observe an additional difficulty: there are simulation-based constraints in Eq. (5.1), which makes the precise determination of the feasibility beforehand extremely difficult. In fact, incorporating constraints efficiently is one of the major challenges in developing any simulation-based optimization methods. One naive and impractical approach is to accurately determine the feasibility of a design (this will be referred to as a perfect feasibility model), then apply OO directly within Θ_f, the subset of all the feasible designs. The other extreme is to apply OO directly regardless of the constraints. This does not work in general since many designs in the selected set may be infeasible. The selected set, the size of which is determined without any consideration of the constraints, can no longer ensure to cover some feasible designs with good enough performance with high probability.

The key idea in this chapter is to note, in practice, although we do not have perfect feasibility model, we usually have some rules, experiences, heuristics, and analytical methods (these will be referred as the feasibility models) to help us find feasible designs with a reasonably high probability (certainly no less than 0.5). These feasibility models usually are not perfect, and some times make mistakes, e.g., some designs may be predicted as feasible by the feasibility model, but are actually infeasible. If we incorporate this fact of imperfect (but with some reasonable chance, say 70% or 80%) feasibility prediction into the determination of the size of the selected set, we can ensure to find some feasible and good enough designs with high probability. Before we discuss how to do this incorporation in details, which will be introduced in Section 1, we would like to make some comments.

Recall that the spirit of OO is to ask for good enough with high probability instead of best for sure. The spirit of the above COO is similar: To accommodate the constraints, we ask for feasibility with high probability instead of feasible for sure. It is interesting to note that the classification of *"feasible* vs. *infeasible"* is **ordinal**. All the advantages of OO apply here, i.e., it can be reasonably easy to obtain a group of truly feasible designs with high probability instead of one for sure. In addition, *"imperfectness"* of the feasibility model is also in tune with the "goal softening" tenet. Although individual determination of feasibility using a crude model may give erroneous results, the model could be very robust with respect to a group of candidates overall. As in the case of the regular OO, $|G \cap S|$ can be good even in the presence of large "noise." The above approach is an evolution of the OO methodology amenable to constrained optimization problems with a *"complete ordinal"* concept. The tenets of *"goal softening"* and *"ordinal comparison"* are reflected by the integration of "imperfect feasibility model" and "feasibility determination."

1 Determination of selected set in COO

As discussed in Chapter III, the effectiveness of the OO technique depends, in part, on the selection rule that we use to select the subset S. The simplest selection rule that requires no performance evaluation or estimate is Blind Pick. Analytical results as shown in Chapter II are available for Blind Pick without constraints as in Eq. (5.2),

$$\text{Prob}\left[|G \cap S| \geq k\right] = \sum_{i=k}^{\min(g,s)} \frac{\binom{g}{i}\binom{N-g}{s-i}}{\binom{N}{s}}. \tag{5.2}$$

Here we will derive analytical results for the constrained cases. When we have constraints, since there are infeasible designs in Θ, if we still use Blind Pick, we will have to select more designs to guarantee the same level of alignment. It is also reasonable to see that the required size of the selected set decreases as the predication accuracy increases.

1.1 Blind pick with an imperfect feasibility model

In engineering practice, we usually have an imperfect feasibility model, which is based on rules, experiences, heuristics, and some analytical methods that can be easily checked. Such a model can make prediction about the feasibility of a design choice θ with little or no computation. However, its prediction will sometimes be faulty. Suppose we first use this feasibility model to obtain N (say 1000) designs from the entire design space Θ as follows. We uniformly sample Θ and test the feasibility with our feasibility model, then accept designs predicted as feasible and reject designs predicted as infeasible. We denote such set of designs as $\hat{\Theta}_f$. Then we apply BP within $\hat{\Theta}_f$ to select a subset S_f. We want to find some truly good enough and feasible deigns of Θ. The rationale here is as follows: when the set of predicted feasible designs $\hat{\Theta}_f$ is large, the density ρ_f of feasible designs in the design space is reasonably high (say no less than 10% of the entire design space)[1], and the feasibility model has a reasonable accuracy (with probability no less than 0.5 to give correct prediction), there should be some truly good enough and feasible designs of Θ contained in $\hat{\Theta}_f$. We call this method Blind Pick with a (imperfect) Feasibility Model (BPFM).

In order to quantify the alignment probability, we denote the set of top-$100 \times \alpha_g\%$ truly feasible designs in $\hat{\Theta}_f$ as G, the good enough set. Suppose there are N_f truly feasible designs in the N predicted feasible designs[2]. Then the size of the good enough set $g = N_f \alpha_g$. We use P_{e1} and P_{e2} to measure the accuracy of a feasibility model, i.e., P_{e1} denotes the probability that a truly feasible design is predicted as infeasible, also known as the type-I error; and P_{e2} denotes the probability that a truly infeasible design is predicted as feasible, also known as the type-II error. To simplify the discussion, let us assume $P_{e1} = P_{e2}$ first, remove this constraint, and discuss the more general case later. Let $P_f = 1 - P_{e1}$, then P_f is the prediction accuracy of the feasibility model, that is, the probability that a design is predicted as feasible if it is truly feasible and also the probability that a design is predicted as infeasible if it is truly infeasible. So, for each θ design in $\hat{\Theta}_f$, which is predicted as feasible, the probability that it is truly feasible can be obtained via Bayesian formula

---

[1] When the density of feasible designs is much less than 10%, we need to improve the value of N or use a good feasibility model so that there are some truly good enough and feasible designs of Θ contained in $\hat{\Theta}_f$.

[2] The selection of N should guarantee N_f is large enough.

$r = \text{Prob}[\theta \text{ is feasible}/\theta \text{ is predicted as feasible}]$

$$= \frac{\text{Prob}[\theta \text{ is feasible and } \theta \text{ is predicted as feasible}]}{\text{Prob}[\theta \text{ is predicted as feasible}]}$$

$$= \frac{\text{Prob}[\theta \text{ is feasible}]\text{Prob}[\theta \text{ is predicted as feasible}/\theta \text{ is feasible}]}{\text{Prob}[\theta \text{ is predicted as feasible}]} \quad (5.3)$$

$$= \frac{\rho_f P_f}{\text{Prob}[\theta \text{ is predicted as feasible}]}$$

$$= \frac{\rho_f P_f}{\rho_f P_f + (1-\rho_f)(1-P_f)}.$$

Thus, because the feasibility model is usually not perfect (that is $P_f<1$), infeasible designs cannot be completely excluded in S_f.

It should be pointed out that although the expected number of feasible designs in $\hat{\Theta}_f$ is $N_f = N \times r$ and the expected number of feasible designs in S_f is $|S_f| \times r$, the results of the regular unconstrained OO method cannot be directly applied in this case with $s = |S_f| \times r$.

Exercise 5.1: Explain intuitively why?

We shall now derive the AP of the selected subset S_f by averaging all possible numbers of feasible designs in S_f.

Suppose the size of selected subset S_f is s_f. The number of infeasible designs in the selected subset S_f, denoted as t_f, follows approximately a *Binomial distribution*, i.e., $t_f \sim b(s_f, r)$, where the size of selected subset s_f is the size of the *Bernoulli trials* and r is the probability that a selected design in S_f is feasible. The probability that there are $t_f = j$ infeasible designs in selected subset S_f, is:

$$\text{Prob}[t_f = j] = \binom{s_f}{j}(r)^{s_f - j}(1-r)^j. \quad (5.4)$$

Given that there are t_f *infeasible designs in the selected subset S_f*, the conditional AP that there are exact k good enough designs in S_f is given by:

$$\text{Prob}\left[|G \cap S_f| = k/t_f\right] = \frac{\binom{g}{k}\binom{N_f - g}{s_f - t_f - k}}{\binom{N_f}{s_f - t_f}}. \quad (5.5)$$

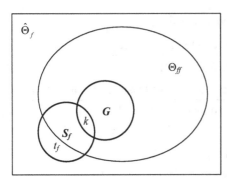

$\hat{\Theta}_f$: the set of predicted feasible designs
Θ_{ff} : the set of truly feasible designs in $\hat{\Theta}_f$
G : the set of good enough designs in $\hat{\Theta}_f$
S_f : the selected set
k : the alignment level
t_f : the number of infeasible designs in S_f

Fig. 5.2. Illustration of Eq. (5.5)

Eq. (5.5) is a direct analog of Eq. (5.2), which was first derived in Eq. (2.37). Please see also Fig. 5.2 for illustration.

Since if there are k feasible and good enough designs in S_f, the number of infeasible designs in S_f could be any number from 0 to $\min(s_f - k, N\text{-}N_f)$, based on the Total-Probability Theorem, we have the formula for the AP that there are at least k good enough designs in the selected set S_f as

$$AP_{COO} = \text{Prob}\Big[\big|G \cap S_f\big| \geq k\Big]$$

$$= \sum_{i=k}^{\min(g,s_f)} \sum_{j=0}^{\min(s_f-i, N-N_f)} \frac{\dbinom{g}{i}\dbinom{N_f - g}{s_f - j - i}}{\dbinom{N_f}{s_f - j}} \dbinom{s_f}{j}(r)^{s_f - j}(1-r)^{j}. \quad (5.6)$$

If we do not have any knowledge about the feasibility of the designs, each of them is equally likely to be feasible or infeasible. This corresponds to the special case where $P_f = 0.5$ and thus $r = \rho_f$ in Eq. (5.6). It is also interesting to note that, if we have perfect knowledge about the feasibility of each design, by sampling only truly feasible designs, we can obtain an $\hat{\Theta}_f$ containing N feasible designs. This corresponds to the special case $P_f = 1$ and thus $r = 1$ in Eq. (5.6). Direct calculation shows that Eq. (5.6) reduces to Eq. (5.2) if $P_f = 1$ and thus $r = 1$.

1.2 Impact of the quality of the feasibility model on BPFM

The value of BPFM lies in the fact that, by only very crude feasibility model, we can bring an impressive improvement to the efficiency of COO. First, we will show that AP_{COO} is an increasing function of P_f, the accuracy of feasibility model and also an increasing function of ρ, the density of feasible designs in the entire design space. We show this result through two steps. In the first step, we show that AP_{COO} is an increasing function of r, and in the second step, we show that r is an increasing function of P_f and ρ_f. Since r represents the probability that an observed feasible design is truly feasible, the number of truly feasible designs on average in $\hat{\Theta}_f$ is Nr, i.e., $N_f = Nr$. Since the good enough set G is defined as the top-100 $\times$ $\alpha_g\%$ of these N_f truly feasible designs, $g = N_f\alpha_g = N\alpha_g r$. Thus, when r increases, g increases. Since we are doing blind picking in $\hat{\Theta}_f$, all other parameters remaining the same (i.e., fixing N and s_f), the AP Prob$[|G \cap S_f| \geq k]$ increases. Now, we show that r is an increasing function of P_f and ρ_f. Fix ρ_f, following Eq. (5.3), we have that

$$\frac{dr}{dP_f} = \frac{\rho_f\left(1-\rho_f\right)}{\left(\rho_f P_f + \left(1-\rho_f\right)\left(1-P_f\right)\right)^2} > 0, \text{ for all } 0 < \rho_f < 1.$$

This shows that r is an increasing function of P_f. Similarly we can show r is an increasing function of ρ_f. In total, we show that the AP Prob$[|G \cap S_f| \geq k]$ is an increasing function of P_f and ρ_f, which is also intuitively reasonable.

Then, we will show some numerical results. Suppose the size of design space $\hat{\Theta}_f$ is 1000. The number of feasible designs is $N_f = 500$. The good enough set G is the top 50 feasible designs (i.e., the top-10% feasible designs in $\hat{\Theta}_f$). The AP versus the size of selected subset, s_f is plotted in Fig. 5.3. As expected, for the constrained problem, the BPFM method with a feasibility model $P_f > 0.5$ is better than that without a feasibility model (i.e., $P_f = 0.50$, which is identical to directly using BP), because, for the same size of the selected subset, the AP obtained for $P_f > 0.50$ is larger than that obtained for $P_f = 0.50$. It is also observed that the more accurate the feasibility model (larger P_f) is, the higher AP we can achieve.

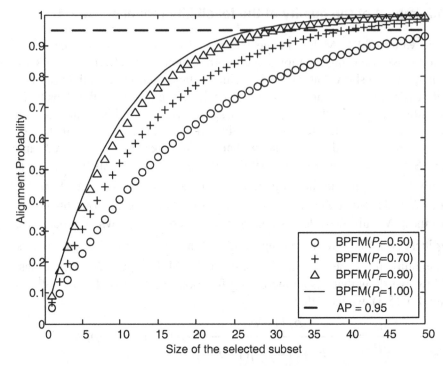

Fig. 5.3. AP versus the subset selection size of BP and BPFM

The sizes of selected subsets obtained by different P_f values are illustrated in Table 5.1. It is shown that, for the same required AP, a larger P_f requires a smaller selected subset, and thus is more efficient. In other words, a smaller selected subset is required for a more accurate feasibility model, for a given level of alignment probability.

Table 5.1. Sizes of the selected subsets

Required AP	$P_f = 0.50$	$P_f = 0.70$	$P_f = 0.90$	$P_f = 1.00$
≥ 0.50	14	10	8	7
≥ 0.60	18	13	10	9
≥ 0.70	24	17	13	12
≥ 0.80	31	22	17	16
≥ 0.90	44	32	24	22
≥ 0.95	57	41	31	28
≥ 0.99	87	61	47	42

So far we assume that $P_{e1} = P_{e2}$ to simplify the notation. Now, we show how to remove this constraint and consider the more general case where r is a function of P_{e1} and P_{e2}. Following a similar analysis to Eq. (5.3), we have

$$r = \frac{\rho_f (1 - P_{e1})}{\rho_f (1 - P_{e1}) + (1 - \rho_f) P_{e2}}. \qquad (5.7)$$

Giving a feasibility model, once we estimate the accuracy of the feasibility model, i.e., P_{e1} and P_{e2}, we can use Eq. (5.7) to calculate r and then use Eq. (5.6) to quantify the AP_{COO}. We now show when the accuracy of the feasibility model increases, i.e., P_{e1} and P_{e2} decreases, r increases, and then following the analysis similar to the beginning of this subsection, we can see that AP also increases. Fix P_{e2} and ρ_f, we have

$$\frac{dr}{dP_{e1}} = -\frac{\rho_f (1 - \rho_f) P_{e2}}{\left(\rho_f (1 - P_{e1}) + (1 - \rho_f) P_{e2} \right)^2} < 0.$$

Similarly, fix P_{e1} and ρ_f, we have

$$\frac{dr}{dP_{e2}} = -\frac{\rho_f (1 - \rho_f)(1 - P_{e1})}{\left(\rho_f (1 - P_{e1}) + (1 - \rho_f) P_{e2} \right)^2} < 0.$$

And fix P_{e1} and P_{e2}, we have

$$\frac{dr}{d\rho_f} = \frac{(1 - P_{e1}) P_{e2}}{\left(\rho_f (1 - P_{e1}) + (1 - \rho_f) P_{e2} \right)^2} > 0.$$

This means r is a decreasing function of P_{e1} and P_{e2}, and an increasing function of ρ_f. The previous discussion on $P_{e1} = P_{e2} = 1 - P_f$ is a special case.

Suppose we are given a feasibility model which predicts the feasibility of a design accurately with probability P_f. We summarize the application procedure of COO using this feasibility model as follows.

Box 5.1. COO approach

Step 1. Find a feasibility model and randomly sample N predicted feasible designs.
Step 2. Specify g and k.
Step 3. Estimate ρ_f, the density of feasible designs in the entire design space and estimate the accuracy of the feasibility model, i.e., the P_{e1} and P_{e2}, and calculate r through Eq. (5.7).
Step 4. Apply the BPFM in Eq. (5.6) to determine the size of the selected set.
Step 5. Randomly select S_f designs from the N designs.
Step 6. The COO theory ensures that there are no less than k good enough feasible designs in the selected subset with high probability.

Exercise 5.2: How can we determine the size of the selected set if we use Horse Race instead of Blind Pick in Step 4 above within the set of predicted feasible designs?

2 Example: Optimization with an imperfect feasibility model

In this section, we use a simple example to evaluate the effects of COO under different observation noise. As expected, since we are developing blind pick based COO, the alignment level of selected set should be insensitive to the level of noise in observation. Let us consider the following constrained optimization problem. Each design θ is an integer between 1 and 1000, i.e., $\Theta=\{1,2,...1000\}$. The objective function $J(\theta) = \theta$. The constraint is that θ must be even numbers. This problem can then be mathematically formulated as

$$\min_{\theta \in \{1,2,...1000\}} J(\theta) = \theta$$
$$\text{s.t. } \text{mod}(\theta,2) = 0 \tag{5.8}$$

where mod($\bullet,\bullet$) is the modulo operator. Suppose also observation noise ξ contains i.i.d. uniform distribution $U(0,a)$ such that our observation is

$$\hat{J}(\theta) \equiv J(\theta) + \xi .$$

The presence of noise makes the optimization problem non-trivial to solve even with perfect knowledge about the feasibility of each design.

Suppose we also have an imperfect feasibility model, which gives the correct feasibility prediction with probability P_f. In other words, when the design θ is an even number (i.e., a truly feasible design), the feasibility model predicts the design as feasible with probability P_f (thus predicts the design as infeasible with probability $P_{e1} = 1\text{-}P_f$); when the design θ is an odd number (i.e., a truly infeasible design), the feasibility model predicts the design as infeasible with probability P_f (thus predicts the design as feasible with probability $P_{e2} = 1\text{-}P_f$). We reasonably assume $P_f \geq 0.5$ (Otherwise we simply reverse the prediction given by this feasibility model, and can then obtain a "reasonable" feasibility model).

Suppose we want to find at least one of the truly top 50 feasible designs with high probability, i.e., $g = 50$, $k = 1$ with $G = \{2,4,6,\dots 100\}$. We simulate the BPFM method with $P_f = 0.95$. Then $P_{e1} = P_{e2} = 0.05$. Notice for this example, half of the designs are feasible, so we have $\rho_f = 0.5$. By Eq. (5.7), we calculate that $r = 0.95$. The size of the selected subset, s_f, then can be calculated based on Eq. (5.6) with the different required AP (0.50, 0.70, and 0.95). The selected subsets S_f for different AP are shown in Table 5.2. It turns out that the BPFM method finds at least one of the good enough feasible designs in all the instances as shown.

Table 5.2. A random examination of BPFM ($P_f = 0.95$)

Required AP	s_f	Selected subset S_f	Alignment level
≥0.50	7	{808, 524, *32*, 850, 240, 498, 878}	1
≥0.70	12	{714, 284, 982, 614, 644, 972, 238, 820, 986, 176, 272, *30*}	1
≥0.90	23	{350, 490, 760, 147, 236, 483, *88*, 130, 260, 456, *24*, 508, 997, 178, 228, 564, 842, 976, 446, 660, 330, 952, 87}	2
≥0.95	30	{842, 812, 716, 682, 980, *8*, 510, 272, 996, 588, 410, 718, 154, 427, 964, 806, 558, 502, 414, 724, 998, 265, 384, 772, 262, 682, 572, 990, 564, 626}	1

Note that the good results in Table 5.2 are not coincidences since the BPFM method blindly picks designs from the space that are predicted as feasible by the feasibility model without relying on accurate estimation on $J(\theta)$. So, we can expect that the guarantee provided by the BPFM method holds no matter how large the noise is.

3 Conclusion

Optimization of DEDS with complicated stochastic constraints is generally very difficult and simulation is usually the only way available. The results on unconstrained OO in Chapter II cannot be applied directly since many infeasible designs cannot be excluded without costly simulation. The COO approach with feasibility model presented in this chapter is effective to solve this long-standing problem. According to No-Free-lunch Theorem (Ho et al. 2003), *no algorithm can do better on the average than blind search without structural information.* The feasibility model in this case can be regarded as the "structural information." As a result, COO provides a more efficient approach for solving constrained optimization problems, since the size of the selected subset is smaller than that when directly applying the unconstrained OO approach.

The algorithm for subset selection and the procedure of *Blind Pick with Feasibility Model* (BPFM) for COO are derived. Numerical testing shows that, by using COO method, to meet the same required alignment probability, *Blind Pick with Feasibility Model* is more efficient than pure Blind Pick. The testing results also show that the method is very robust, even when the feasibility model is not very accurate. Furthermore, the COO method presented in this chapter is a general approach. Any crude feasibility model even with large noise is compatible and can work well with the approach. In Chapter VIII Section 3, we apply COO with a feasibility model based on the rough set theory to a real world remanufacturing system, and yields promising results. ***Similarly, the application of this approach of COO is not restricted to the BP selection method. Other selection methods such as the Horse Race method can also be used in connection with the crude feasibility model. The modifications to the AP of course must be carried out similar to that of Eq. (5.6) except via simulation.*** A quick-and-dirty first approximation is to simply modify the unconstrained UAP by r.

Chapter VI Memory Limited Strategy Optimization

Let us start with a big picture to describe the relationship between this chapter and the previous chapters. We focus on how to solve a simulation-based strategy optimization problem[1]. Conceptually, we need three components: a program which implements a strategy γ, a performance evaluation technique to calculate $J(\gamma)$ when γ is applied to the system, and an optimization algorithm to find the optimal or good enough strategy. The relationship among these three components is shown in Fig. 6.1.

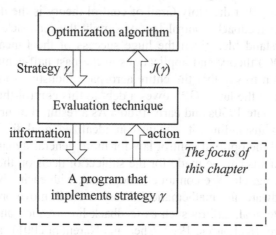

Fig. 6.1. A big picture for simulation-based strategy optimization

Note that the optimization algorithm only requires the evaluation technique to return the performance of a strategy, but does not care about how this evaluation is obtained; the evaluation technique only requires the program to return the corresponding action of a strategy when the information is the input. When implementation is considered, we have to make

---

[1] A strategy is simply a mapping from input information to output decision or control. Strategy is also known as decision rule, if-then table, fuzzy logic, learning and adaptation algorithm, and host of other names. However, nothing can be more general than the definition of a function that maps all available information into decision or action.

sure the strategy is simple enough so that it can be stored within given limited memory, in other words, we require a strategy is simple enough to be implementable. Furthermore, when ordinal optimization is considered, one challenge will be sampling of design space. Sampling is easy when we have a natural representation of a design as a number or a vector as we have done for previous chapters. In the context of strategy optimization, representing all implementable strategies so that sampling can be taken seems nontrivial. The focus of this chapter is to provide a systematic representation of strategy space so that the optimization algorithms developed earlier can be applied.

1 Motivation (the need to find good enough and simple strategies)

It can be argued that the Holy Grail of control theory is the determination of the optimal feedback control law or simply the feedback control law. This is understandable, given the huge success of the Linear-Quadratic-Gaussian (LQG) theory and applications in the past half-century. It is not an exaggeration to say that the entire aerospace industry from the Apollo moon landing to the latest GPS owes a debt to this control-theoretic development in the late 1950s and early 1960s. As a result, the curse of dimensionality notwithstanding, it remains an idealized goal for all problem solvers to find the optimal control law for more general dynamic systems. Similar statements can be made for the subject of decision theory, adaptation and learning, etc. We continue to hope that, with each advancement in computer hardware and mathematical theory; we will move one step closer to this ultimate goal. Efforts such as feedback linearization and multimode adaptive control (Kokotovic 1992; Chen and Narendra 2001) can be viewed as such successful attempts.

The theme of this chapter is to argue that this idealized goal of control theory is somewhat misplaced. We have been seduced by our early successes with the LQG theory and its extensions. There is a simple but always neglected fact that *it is extremely difficult to specify and impossible to implement a general multivariable function even if the function is known.*

Generally speaking, a one variable function is a two-column table; a two-variable function is then a book of tables; a three-variable function, a library of books; four-variable, a universe of libraries; and so on. Thus, how can one store or specify a general arbitrary 100-variable function never mind implementing it even if the function is God given? No hardware advancement will overcome this fundamental impossibility, even if mathematical advancements provide the general solution. This is also clear from the following simple calculation. Suppose there are n-bit input information and m-bit output action for a strategy. To describe such a strategy as a lookup table, we need to store all the (information, action) pairs. There are 2^n such pairs in total, and we need $(n + m)$ bits to store each pair. Thus we need $(n + m)2^n$ bits to store a strategy. When $n = 100$, $m = 1$, this number is 101×2^{100} bits $\approx 2^{107}$ bits $= 2^{74}$ Gega Bytes (GB), which exceeds the memory space of any digital computer known nowadays or the foreseeable future. Exponential growth is one law that cannot be overcome in general. Our earlier successes with the Linear-Quadratic-Guassian control theory and its extensions are enabled by the fact that the functions involved have very a special form, namely, they decompose into sums or products of functions of single variable or low dimensions. As we move from the control of continuous variable dynamic systems to discrete event systems or the more complex human-made systems discussed in this book, there is no prior reason to expect that the optimal control law for such system will have the convenient additive or multiplicative form. Even if in the unlikely scenario that we are lucky to have such simple functional form for the control law of the systems under study, our efforts should be to concentrate on searches for actual implementation of such systems, as oppose to finding the more general form of control law.

In this light, it is not surprising that "Divide and Conquer" or "Hierarchy" is a time-tested method that has successfully evolved over human history to tackle many complex problems. It is the only known antidote to exponential growth. Furthermore, by breaking down a large problem into ever-smaller problems, many useful tools that do not scale up well can be used on these smaller problems. Decomposing a multivariable function into a weighted sum of one-variable functions is a simple example of this principle. In addition, Nature has also appreciated this fundamental difficulty of multivariable dependence. There are many examples of adaptation, using simple strategies based on local information and neighboring interactions to achieve successful global results abound (Think globally but act locally), such as ants, bees, germs, and viruses (Vertosick 2002). Recent research on the No-Free-Lunch theorem (Ho and Pepyne 2004) also points to the importance and feasibility of "simple" control laws for complex systems. And as we venture into the sociological and psychological realm, there are even more evidences showing that it only leads to unhappiness and non-optimality to strive for the "best" (Schwartz 2004).

The purpose of this chapter is to discuss systematic and computationally feasible ways to find "good enough" AND "simple" strategies. Since we will focus on simulation-based strategy optimization, many difficulties mentioned in earlier chapters remain, such as the time-consuming performance evaluation and the large design space. In addition we have one more difficulty, that is the constraint on the limited memory space to store strategies.

2 Good enough simple strategy search based on OO

2.1 Building crude model

It is important to understand that lookup table or brute force storage and representation is usually not an efficient way to implement a strategy and is infeasible and impractical for almost all large scale problems. Recall that a strategy is a mapping from the information space to the action space. In other words, a strategy determines what to do when specific information is obtained. As long as we find a clever way (such as using a program) to generate the output for any given input, we can represent the strategy. The size of memory we use may be much less than the lookup table. To identify simple strategies (or strategies that need less memory than a certain given limit), we need to introduce the concept of descriptive complexity (also known as the Kolmogorov complexity (KC) (Li and Vitányi 1997)) which mathematically quantifies the minimal memory space that is needed to store a function. A. N. Kolmogorov developed this concept in 1965 (Kolmogorov 1965). The Kolmogorov complexity of a binary string s is defined as the length of the shortest program for a given universal Turing machine U (explanation follows) to output the string, i.e.,

$$C_U(s) = \min_p \{|p| : \psi_U(p) = s\},$$

where $\psi_U(p)$ represents the output of the universal Turing machine U, when program p is executed on U. Roughly speaking, a universal Turing machine (UTM) is a mathematical model of the computers we are using nowadays, which consists of the hardware (e.g., the hard drive and the memory chip to store the program, and equipment to read from and write on the hard drive and the memory chip) and the software (including low level system software such as operating systems, e.g., Microsoft Windows and Mac OS, and application software p developed for a specific task). Obviously KC depends on which UTM is used. This is reasonable and practical when we use computers to search for a good simple strategy, the hardware and the software in the computer are already given, so the U is fixed. In the following discussion, we will omit the subscript U, when there is no confusion. Giving the concept of KC, in principle, we can judge whether the KC of a strategy is within the given memory limit. Using the terminology of OO, KC is the true model to determine whether a strategy is simple. Unfortunately, it is a basic result in the theory of Kolmogorov complexity that the KC cannot be computed by any computers precisely in general (Theorem 2.3.2, p. 121, (Li and Vitányi 1997)). From an engineering viewpoint, this means that it is extremely time-consuming to find out the true Kolmogorov performance of the proposed strategy, if it is not impossible. Thus, the methodology of OO naturally leads us to consider the usage of approximation, which is computationally fast to replace it[2,3], and to sample simple strategies. In the rest of this chapter, we will formulate this idea of simple strategy generation based on estimated descriptive complexity, which can then be utilized even if the user has little knowledge or experience of what a simple strategy might look like. This is in contrast to existing efforts where no quantification on the descriptive complexity for the strategies is explored. Examples include threshold type of strategies, Neurodynamic programming (NDP) (Bertsekas and Tsitsiklis 1996) which uses neural networks to parameterize strategy space, State aggregation (Ren and Krogh 2002), time aggregation (Cao et al. 2002), action aggregation (Xia et al. 2004), and event-based optimization (Cao 2005).

The crude model of the KC for a strategy we would like to introduce here is the size of a program based on the reduced ordered binary decision diagram (ROBDD, or simply OBDD) representation for the strategy which will be introduced below. ROBDD regards each strategy as a (high-dimensional) Boolean function. (For simplicity we let the output decision variable be binary. This can be generalized in obvious ways. See Exercises 6.1 and 6.2 below.) The observation behind this is that reduced ordered binary decision diagrams (ROBDDs) usually supply a succinct description for a Boolean function (Andersen 1997). Let us first describe how ROBDD can be obtained for a Boolean function and furthermore for a strategy through an example.

---

[2] In the same spirit of OO with constraints in Chapter V but different in that we are using an upper bound estimation for memory used for describing a strategy, so we never include infeasible strategies in our selected set, nor do we know when a strategy is not estimated as simple, what is the probability for it to be truly simple.

[3] It should be noted that although KC in general cannot be calculated by computers, there are extensions of KC that can be calculated by computers, such as the Levin complexity (Levin 1973, 1984), which considers both the length of the program and the time for the program to be executed. It is still an open question how to combine Levin complexity with ordinal optimization to find simple and good enough strategies.

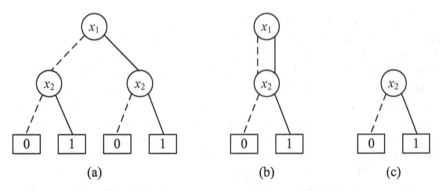

Fig. 6.2. Reduction process from BDD to OBDD for the function $f(x_1,x_2) = (x_1 \wedge x_2) \vee (\neg x_1 \wedge x_2)$

In Fig. 6.2(a), we use a BDD (Binary Decision Diagram) to describe the Boolean function $f(x_1,x_2) = (x_1 \wedge x_2) \vee (\neg x_1 \wedge x_2)$, where $\wedge$ is AND, $\vee$ is OR, and $\neg$ is NOT. To construct a BDD for f, we start from a root node (circle) representing a Boolean variable (x_1 for this example). We connect to this node a dotted line and a solid line representing "if the variable takes value "0" or "1", respectively. For each branch (dotted line or solid line), we add a new node (circle) by choosing a new Boolean variable (x_2 for this example). We keep this branching procedure until all Boolean variables have been added and two lines added. Note that any path from the root node to a latest added node is corresponding to an assignment of 0 or 1 to all Boolean variables. For example, in Fig. 6.2(a), the path in which both node x_1 and x_2 take the dotted branch is corresponding to the assignment $(x_1, x_2) = (0,0)$. As the last step to construct a BDD, we add at the end of each path a box labeled by the evaluation of f under the assignment corresponding to the path. For the assignment $(x_1, x_2) = (0,0)$, we attach a box labeled "0" because $f(0,0) = 0$. Before doing any reduction, BDD will have an exponentially (in n) large number of nodes. One way to reduce BDD is to introduce order when adding Boolean variables. If, in a BDD, all paths choose the same order when adding new Boolean variables, we get an OBDD, where the first "O" stands for "ordered". Fig. 6.2(a) is in fact an OBDD.

OBDDs allow us to find simpler representation for Boolean functions. We can combine the redundant nodes, i.e., the nodes with identical subgraphs. For example, Fig. 6.2(b) gives a more compact OBDD than Fig. 6.2(a). Obviously, if both of the two lines connected to a node are connected to the same successor on the other end (e.g., the lines connected to node x_1 in Fig. 6.2(b)), this means the input value of this Boolean variable does not affect the output of the function. So this node can be removed to make the OBDD more compact. The OBDD in Fig. 6.2(b) can be further simplified to the one in Fig. 6.2(c), where node x_1 is removed. By eliminating redundancies in an OBDD, a unique OBDD can be obtained which is called ROBDD[4] for the Boolean function. In the rest, when we mention an OBDD of a strategy, we will always refer to the ROBDD for the strategy.

Exercise 6.1: How can we encode a strategy from a finite information space to a finite action space with a high dimensional Boolean function?

The readers may consult Chapter VIII. 4 for such an example.

Exercise 6.2: How can we generalize the above OBDD to represent more than one-bit outputs, say two bits? In other words, there are totally four actions, 00, 01, 10, and 11.

Exercise 6.3: Currently there is no randomness in OBDDs. Is it possible to introduce any randomness in OBDDs? In other words, instead of deterministically selecting either the dotted line or the solid line and thus deterministically outputting 0 or 1 finally, can we generalize the OBDD to randomly output 0 or 1? How? If possible, please show the example when the OBDD outputs two-bit actions. What is the advantage of these random OBDDs comparing with the deterministic OBDDs?

Once we have an OBDD for a Boolean function describing a strategy, we can follow a natural way to convert the OBDD to a program that can represent the strategy. For a given input to the strategy, the purpose of the program is to generate the output (either 0 or 1 if the strategy has only two actions to choose) for the strategy. We start from the top node of the OBDD, considering which line to choose (and thus which successor to go to) according to the input values of the Boolean variables until arriving at the bottom box (either 0 box or 1 box), and then output the value in the box. This procedure can be described by a sequence of rules. Each rule looks like

$$(state, input, action, next\ state),$$

where *state* represents which node the program is currently at, *input* represents the input value of the Boolean variable associated with that node, *action* describes what the program is going to do (such as to choose

---

[4] The ROBDD depends on the order of variables.

either of the lines if the program is staying at a node; or to output either "0" or "1", if the program is at one of the bottom boxes; or simply to end the program if the output is already done), and the *next state* represents the node that the program is going to (either the low- or the high-successor of the current node, if the program is now staying at a node; or an END state which describes the end of the program, if the program is now staying at one of the bottom boxes.). For example, the rules to describe the OBDD in Fig. 6.2(c) are:

(node x_2, 0, choose the dotted line, box 0),
(node x_2, 1, choose the solid line, box 1),
(box 0, ә, output 0, END),
(box 1, ә, output 1, END),

where ә means that no input is needed.

Based on this program representation of a strategy, we can estimate its KC as $\hat{C}(\gamma) = 4(2b + 2)\lceil \log_2(b+3+4)\rceil$ by calculating number of bits to implement the strategy, where b is the number of nodes (excluding the bottom boxes) of the OBDD and $\lceil a \rceil$ represents the minimal integer no less than a. In fact, we have the following observations. In general, there are 2 rules associated with each of the nodes (excluding the bottom boxes), and there is a rule associated with each bottom box. Then the number of rules is $r = 2b + 2$. To describe each such rule, we need to encode each of the four elements in a rule by binary sequences. Since we need to distinguish all the b nodes, 2 bottom boxes, the END state, and the 4 possible actions to take (choose either the dotted line or the solid line, output either 0 or 1), we need $d = \lceil \log_2(b+3+4)\rceil$ bits to describe each element. Thus, in total, we need $4rd$ bits to implement an OBDD. Note that $4rd$ is only an estimate on the minimal number of bits to describe an OBDD. First, different order in Boolean variables may lead to OBDD with different size. Unfortunately it is too time-consuming to find the simplest OBDD to describe a strategy in general (which has been proven to be NP-hard (Bollig and Wegener 1996)). Second, there may be different requirements on the rules in different computer systems. For example, some computer systems may allow us to encode four elements separately, which means the computer knows which one of the four elements it is reading, then we can further save the number of bits to represent a rule. In some other computer systems, the value of r and d are required to be clearly explained to the computer. r and d need to be encoded in specific ways to ensure the computer understands them. Considering the different requirements in different computer systems, we may have a more detailed and more specific estimate of the number of bits to represent a strategy γ. Examples can be found in (Jia et al. 2006b).

In summary, to simplify the discussion, we use $\hat{C}(\gamma)$ to represent the number of bits given by whatever approximation. The users are free to use either $\hat{C}(\gamma) = 4rd$ or any other problem-specific estimates.

Exercise 6.4: How can we modify $\hat{C}(\gamma) = 4rd$ when there are m-bit outputs?

2.2 Random sampling in the design space of simple strategies

Once we have a way to estimate the descriptive complexity (KC) for a strategy as above, to take advantage of OO in searching a small set of strategies that contains given number of good enough simple strategies with high probability, we have to find a way to do random sampling in the set of strategies describable within the given memory limit[5]. Our idea is to sample only the estimated simple strategies. More specifically, we randomly generate OBDDs so that the estimated number of bits to describe this OBDD does not exceed the given memory space C_0, i.e., $\hat{C}(\gamma) \leq C_0$. The strategy described by this OBDD is by definition an estimated simple strategy[6]. By sampling these OBDDs, we are sampling simple strategies. One question of this is, as we explained earlier, there might be several OBDDs representing the same strategy, uniformly sampling the OBDDs

---

[5] Note in general, it is impossible to enumerate all simple strategies since the total number of simple strategies is still large.

[6] Since we are using estimation, some truly simple strategies may be excluded. Some readers might be curious to know how many true simple strategies may be excluded. Honestly, this is a difficult question. One reason is that this difference depends on which UTM is used, i.e., the hardware and the software in the computer that we use to do the optimization. Although the difference between the KC of a given string s in different UTM can be bounded by a constant, which is independent from s and only depends on the two UTMs (Li and Vitányi 1997), this constant might be large. This means the same estimate of KC might exclude different numbers of true simple strategies when different UTMs are used. However, how to choose the UTM, i.e., which software or hardware to use, is also an optimization problem, which is probably not easy. It is still an open question to study how many true simple strategies are excluded by a given estimate of KC. Thus, ultimately we must still let the end result justify our approach. See Chapter VIII for an example.

might not mean uniformly sampling the simple strategies. After introducing some restrictions, say we fix the order of the variables from the top node to the bottom box, and combine all the redundant nodes, the sampling redundancy can be sufficiently reduced. To distinguish from the usual OBDD, we call such an OBDD a partially reduced OBDD (PROBDD).

The definition of PROBDD ensures the uniqueness of the nodes in each level (the top node is in level 1 and there are at most n levels), which allows us to say: no two PROBDDs with the same number of levels represent the same Boolean function. Astute reader might notice that although n-level PROBDDs can represent all the 2^{2^n} strategies using n-bit information, some strategies that do not use all the n bit information can be represented by simpler PROBDDs. However, since there are 2^{2^i} different i-level PROBDDs, and all the Boolean functions are represented by an i-level PROBDD (where the order of the variables is $x_1...x_i$) can be represented by exactly an $(i+1)$-level PROBDD (where the order of the variables is $x_1...x_i x_{i+1}$), among all the 2^{2^n} strategies using n-bit information, 2^{2^i} strategies can be represented by i-level PROBDDs, $i = 1,2,...n$. This result brings us two advantages. First, suppose we start from 1-level PROBDDs and incrementally increase the number of levels, until we generate all the 2^{2^n} strategies. We generate at most $\sum_{i=1}^{n} 2^{2^i}$ PROBDDs in total. The redundancy is

$$\sum_{i=1}^{n-1} 2^{2^i} / 2^{2^n} \approx 2^{-2^{n-1}},$$

which reduces to zero faster than an exponent when n increases. This shows the high efficiency of the aforementioned sampling method of simple strategies. The redundancy is ignorable. As an example, for $n = 1,2,3,$ and 4, we test the redundancy numerically and show in Table 6.1, where the Redundancy = (Total PROBDD # – Total Strategy #)/Total Strategy # × 100%. For n = 4, the redundancy has already been very small (less than 1%). The implication is that, for large n and a given memory space, it is sufficient to uniformly sample PROBDDs for obtaining uniform samples from the estimated simple strategy space defined by $\left\{ \gamma : \hat{C}(\gamma) \le C_0 \right\}$.

Table 6.1. The small redundancy of the sampling method of simple strategies (Jia et al. 2006b) © 2006 IEEE

n	Total Strategy # (2^{2^n})	Total PROBDD #	Redundancy (%)
1	4	4	0
2	16	18	12.5
3	256	272	6.25
4	65536	65806	0.412

To uniformly sample PROBDDs, we first fix the order of the variables in all levels of the PROBDD, say $x_1, x_2, ...x_n$. Then we estimate what is the largest number of nodes that can be stored in the given memory space, denoted as b_{max}. We randomly pick an integer b between 0 and b_{max}, where 0 means that the PROBDD does not use any input information and always outputs 0 (or 1). Based on b, we then determine the number of the levels in the PROBDD and the number of nodes in each PROBDD. After that we randomly determine the type of the connections between the nodes in two adjacent levels (including the connections between the nodes in the last level and the bottom boxes), i.e., whether a line between two nodes is dotted or solid. In this way, we can randomly generate a PROBDD that is estimated simple.

Recall the big picture in Fig. 6.1. Once the PROBDDs representing simple strategies are randomly sampled, we remove the constraint on limited memory space from the original simulation-based strategy optimization problem. In the OO procedure, this means we have the N sampled designs from the entire design space now, i.e., Θ_N. Then we can use standard OO to find strategies in Θ_N with good enough performances as described in Chapter II. In this way, we can find simple and good enough strategies with high probability. We show an example to illustrate this procedure in details in Section VIII.4.2.

Exercise 6.5: Besides saving the memory space, what are the other advantages of simple strategies?

3 Conclusion

In summary, this chapter discusses the importance of considering the constraint of limited memory space when applying computer-based control and optimization in large scale simulation-based strategy optimization. This constraint is one of the important reasons why we can only search within the simple strategies in practice. We use multivariate Boolean functions to represent a strategy. OBDD is an efficient conceptual way to represent n-variable Boolean functions. We have developed a method to systematically explore the n-variable Boolean functions that can be captured by i-variable ($i<n$) Boolean functions for $i = 1,2,\ldots$. This exploration can be easily combined with OO to find a strategy with good enough performance and i-variable Boolean function representation for an optimization problem. In Chapter VIII Section 4, we demonstrate this on the well known Witsenhausen problem and obtain a 40-fold decrease in strategy complexity with minor (within 5%) degradation of performance.

Chapter VII Additional Extensions of the OO Methodology

In this chapter we will discuss some other extensions and related issues of ordinal optimization. First, in Section II.5, we propose to use a random sample of N (e.g., $N = 1000$) designs as a representative of the entire design space, and then use the UAP table to determine the selected size such that there are some truly good enough designs included in the selected set. When the design space is extremely large, astute readers may ask whether we can find some truly good enough designs from the entire design space by looking at only these N designs. Roughly speaking, the answer is positive for a reasonably large good enough set (say we look for some top-5% designs of the entire design space). We discuss this issue in Section 1. Second, philosophically, OO allows parallel performance evaluation of the randomly sampled N designs. While a massively parallel computer can be used for this in an obvious way, the process can also be carried out very efficiently on a regular non-parallel computer, provided we are willing to assume some acceptable approximations. The idea is to share some considerable portion of the simulation for structurally similar but parametrically different systems. For more details, please refer to Section 2. Third, in Chapter II, we assume the observation noise is i.i.d.. In practice, sometimes the observation noises are correlated. Can OO still work in this case? The answer is again positive. Actually, as will be shown in Section 3, the correlation among the observation noises seldom hurts and usually helps in OO, i.e., to require a smaller selected set. Fourth, due to the broad practical applications, we use a separate section (Section 4) to introduce the optimal computing budget allocation (OCBA) which serves as a specific selection rule of OO, and the nested partition (NP) which serves as a possible framework to do OO iteratively. Fifth, in conventional OO, the good enough set is defined as the top-n% designs, i.e., based on the ordinal performances. A natural question is whether we can define the good enough set based on the cardinal performances? Can we apply OO in this case? Though the answer to these questions are not complete yet, in Section 5, we try to share some ideas on how to apply OO when the good enough set is defined based on cardinal values rather than ordinal performances.

Finally in Section 6, we discuss the combination of OO and other optimization algorithms, such as genetic algorithm, simulated annealing, tabu search, and Lagrangian relaxation, just to name a few.

1 Extremely large design space

This section is mainly based on the work of S. Y. Lin and Y. C. Ho in (Lin and Ho 2002). One important contribution of OO is to allow us to use a crude model to discover some truly good enough designs with high probability, e.g., Prob$[|G_\Theta \cap S| \geq k] \geq 0.95$, where G_Θ denotes the top-n% designs in the entire design space Θ. Recall the application procedure introduced in Chapter II. We first randomly sample a large number of N designs (usually $N = 1000$) from the entire design space Θ, and then apply OO to screen out some observed good designs to cover some truly good enough designs with high probability. There is an implicit assumption here, i.e., the set of these N designs (Θ_N) is a reasonable representative of the entire design space. Under this assumption, when we find some truly top-n% designs of Θ_N, it is natural to believe that we also find some truly good enough designs of the entire design space Θ. Astute readers may realize this assumption might not be true in some cases. In Fig. 7.1 we visualize the difference between the two good enough sets, where G_Θ and G_N denote the truly top-n% designs in Θ and Θ_N, respectively. Because these N designs are randomly sampled, there are chances that $G_N \not\subseteq G_\Theta$. When this happens, it is obvious that

$$\text{Prob}\left[|G_\Theta \cap S| \geq k / G_N \not\subseteq G_\Theta\right] < \text{Prob}\left[|G_N \cap S| \geq k / G_N \not\subseteq G_\Theta\right],$$

then the selected set S may not contain some designs in G_Θ with high enough probability. To justify the application of OO in an extremely large design space, it is inevitable to ask the question: How can we select S such that some truly good enough designs of the entire design space are contained with high probability, i.e., Prob$[|G_\Theta \cap S| \geq k]$ is high? This is just what we will answer in this section. Before we go to detailed discussion, let us present the answer first: *As we will show that the two alignment probabilities are very close to each other, we can treat Θ_N with $N \geq 1000$ as a reasonable representative of Θ and apply the UAP table in Section II.5 to determine the selected*

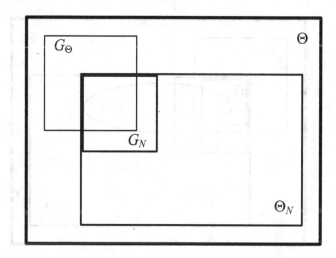

Fig. 7.1. It is possible that $G_N \not\subseteq G_\Theta$

size of S, with little concern for ordinary applications[1]. Practitioners can now skip the rest of this section and go ahead applying OO with little worry. For readers interested in this, let us see how the above answer is obtained.

In OO, what we can control is how to select S such that $\text{Prob}[|G_N \cap S| \geq k]$ is high for the given top-n% good enough designs in Θ_N. However, some designs in G_N may not rank top-n% in the entire design space Θ. It is important to understand how many truly good designs G_Θ, say top-n% designs in Θ, are covered by S with high probability. For OO to be usable, this overlap level should be high. This turns out to be true, as will be explained in detail below.

The basic idea is to show that for a slightly reduced good enough set, denoted as G_N^r, say top-m% designs of Θ_N, where $m/n = 0.7$, it is possible to guarantee $G_N^r \subseteq G_\Theta$ with probability near 1. In Fig. 7.2, we illustrate such a situation. Since m is less than n, the price we have to pay is that the alignment level between S and G_Θ is also less than k. Let us denote it as k'.

---

[1] In US presidential elections with over 100 million voters, one can predict the outcome based on the exit voting interview with some 1000 typical voters. Intuitively, we believe the same idea can be applied to determine good or bad designs so long as the ratio of $|G|/|\Theta|$ is not too small.

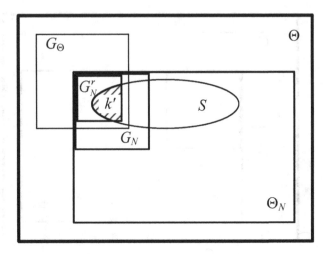

Fig. 7.2. It is possible that $G_N^r \subseteq G_\Theta$

Now we proceed to find a lower bound for the alignment probability $\text{Prob}\left[\left|G_\Theta \cap S\right| \geq k'\right]$ and show it is high enough for k' slightly smaller than k. Let S_k be the set of the truly top-k designs in S then, $S_k \subset S$. Thus we have

$$\text{Prob}\left[\left|G_\Theta \cap S\right| \geq k'\right] \geq \text{Prob}\left[\left|G_N \cap S\right| \geq k, \left|G_N^r \cap S_k\right| \geq k', G_N^r \subseteq G_\Theta\right]. \quad (7.1)$$

since the joint event at the right side is only a special case for $\{\left|G_\Theta \cap S\right| \geq k'\}$ to occur. We can rewrite the right hand of Eq. (7.1) as

$$\text{Prob}\left[\left|G_N \cap S\right| \geq k, \left|G_N^r \cap S_k\right| \geq k', G_N^r \subseteq G_\Theta\right]$$
$$=\text{Prob}\left[\left|G_N \cap S\right| \geq k\right]\text{Prob}\left[\left|G_N^r \cap S_k\right| \geq k', G_N^r \subseteq G_\Theta \middle/ \left|G_N \cap S\right| \geq k\right] \quad (7.2)$$
$$=\text{Prob}\left[\left|G_N \cap S\right| \geq k\right]\text{Prob}\left[G_N^r \subseteq G_\Theta \middle/ \left|G_N \cap S\right| \geq k\right]$$
$$\times \text{Prob}\left[\left|G_N^r \cap S_k\right| \geq k' \middle/ G_N^r \subseteq G_\Theta, \left|G_N \cap S\right| \geq k\right].$$

Because the event $\{\left|G_N \cap S\right| \geq k\}$ has nothing to do with the event $\{G_N^r \subseteq G_\Theta\}$, these two events are independent, i.e.,

$$\text{Prob}\left[G_N^r \subseteq G_\Theta / |G_N \cap S| \geq k\right] = \text{Prob}\left[G_N^r \subseteq G_\Theta\right]. \tag{7.3}$$

Also the event $\left\{|G_N^r \cap S_k| \geq k'\right\}$ and the event $\left\{G_N^r \subseteq G_\Theta\right\}$ are independent, so

$$\text{Prob}\left[|G_N^r \cap S_k| \geq k' / G_N^r \subseteq G_\Theta, |G_N \cap S| \geq k\right]$$
$$= \text{Prob}\left[|G_N^r \cap S_k| \geq k' / |G_N \cap S| \geq k\right]. \tag{7.4}$$

Combining Eqs. (7.1)-(7.4) together, we have

$$\text{Prob}\left[|G_\Theta \cap S| \geq k'\right]$$
$$\geq \text{Prob}\left[|G_N \cap S| \geq k\right] \text{Prob}\left[G_N^r \subseteq G_\Theta\right] \tag{7.5}$$
$$\times \text{Prob}\left[|G_N^r \cap S_k| \geq k' / |G_N \cap S| \geq k\right].$$

The first term in the right hand of Eq. (7.5) is already dealt with by the UAP table in Section II. 5. We shall show later that both the second and the third terms are near 1 for reasonable values of m and k'. For the second term, it is obvious that

$$\text{Prob}\left[G_N^r \subseteq G_\Theta\right] = 1 - \sum_{i=0}^{|G_N^r|-1} \text{Prob}\left[|G_\Theta \cap \Theta_N| = i\right], \tag{7.6}$$

where $\text{Prob}[|G_\Theta \cap \Theta_N| = i]$ is the probability that exactly i good enough designs of the entire design space Θ are contained in the set of N randomly sampled designs. Because these N designs are assumed to be uniformly sampled from the entire design space, we have

$$\text{Prob}\left[|G_\Theta \cap \Theta_N| = i\right] = \frac{\binom{|G_\Theta|}{i}\binom{|\Theta| - |G_\Theta|}{N-i}}{\binom{|\Theta|}{N}}, \tag{7.7}$$

which looks similar to the formula used in the Blind Pick rule in Chapter II. Since the design space Θ is usually extremely large, say no less than

10^8, each time when a design is uniformly sampled from Θ, we approximately have probability $|G_\Theta|/|\Theta|$ to sample a design inside G_Θ, and have probability $1-|G_\Theta|/|\Theta|$ to sample a design outside G_Θ. In a sequence of N samples, under the condition that there are exactly i samples from G_Θ, there are $\binom{N}{i}$ possible combinations for these i samples to appear in the sampling process. So, Eq. (7.7) can be approximated by

$$\mathrm{Prob}\left[\left|G_\Theta \cap \Theta_N\right|=i\right]\approx\binom{N}{i}\times\left(\frac{|G_\Theta|}{|\Theta|}\right)^i\times\left(1-\frac{|G_\Theta|}{|\Theta|}\right)^{N-i}. \qquad (7.8)$$

So the second term in the right hand of Eq. (7.5) can be calculated by Eq. (7.6) and Eq. (7.8). For the third term, researchers have done abundant experiments to study the relationship between the values of k and k' s.t. $\mathrm{Prob}\left[\left|G_N^r \cap S_k\right|\geq k'\big/\left|G_N \cap S\right|\geq k\right]$ is close to 1. For example, when $N=1000$, G_N and G_N^r represent the top-5% and top-3.5% designs in Θ_N, respectively (i.e., $n=5$, $m/n=0.70$), if k and k' take the values in Table 7.1, then $\mathrm{Prob}\left[\left|G_N^r \cap S_k\right|\geq k'\big/\left|G_N \cap S\right|\geq k\right]=1$.

Table 7.1. Relationship between k and k' for $\mathrm{Prob}\left[\left|G_N^r \cap S_k\right|\geq k'\big/\left|G_N \cap S\right|\geq k\right]=1$

k	k'
1	1
2	1
3	2
4	3
5	4
6	4
7	5
8	6
9	7
10	7

Now we can easily calculate the right hand of Eq. (7.5). To get a rough idea how close the two alignment probabilities $\mathrm{Prob}[|G_\Theta \cap S|\geq k']$ and $\mathrm{Prob}[|G_N \cap S|\geq k]$ are, we show one instance as follows. For $|\Theta|=10^8$, $N=1000$, $m/n=0.70$, $n=5$, then from Eq. (7.6) and (7.8), we have

$\text{Prob}\left[G_N^r \subseteq G_\Theta\right] = 0.991$. If $\text{Prob}[|G_N \cap S| \geq k] = 0.95$ and the values of k and k' are chosen s.t. $\text{Prob}\left[\left|G_N^r \cap S_k\right| \geq k' / \left|G_N \cap S\right| \geq k\right] = 1$, then Prob $[|G_\Theta \cap S| \geq k'] \geq 0.942$. The two probabilities are very close to each other. This means if we follow the UAP table in Section II.5 to determine the size of the set S, then we have $\text{Prob}[|G_N \cap S| \geq k] \geq 0.95$ and $\text{Prob}[|G_\Theta \cap S| \geq k'] \geq 0.942$. Since all we want in engineering applications is a high probability, the little difference between the two alignment probabilities can be ignored.

In summary, we show in this section that $N \geq 1000$ is usually a reasonable representative of the large design space Θ. Practitioners can go ahead applying the OO introduced in Chapter II to find some good enough designs of the entire design space with little worry. The bottom line is: If we choose a more restricted definition of G_N about 30% less than G_Θ, i.e., $m/n = 0.70$, we can use the UAP table in Section II.5 to predict the size of set S such that $\text{Prob}[|G_\Theta \cap S| \geq k]$ is high.

On the other hand, we cannot push this line of reasoning too far. If $|\Theta|$ is sufficiently large, say 10^9, and we require $G = \text{top } 20$. Simple analysis will show that the probability a uniform sample of $N = 1000$ will have little chance of containing a single true top-20 design.

Exercise 7.1: Prove this and by the same analysis discuss how large G or $g/|\Theta|$ must be for the results of this section to be valid.

2 Parallel implementation of OO

Many optimization problems involve search in a multi-dimensional space of parameters. In a simulation program of such problems, the structure of the computational instructions does not change when the parameters change. In the parlance of parallel computation, this is called Single Instruction and Multiple Data (SIMD) mode. Since OO typically involves the evaluation of N (e.g., $N = 1000$) estimated performances using a crude model, an efficient simulation for doing this can be implemented without separately running N simulations each using a different parameter combination. While a massively parallel computer can be used for this in an obvious way, we want to show here that the process can also be carried out on a regular non-parallel computer very efficiently, provided we are willing to assume some acceptable approximations. The Standard Clock (SC) method that will be introduced later in this section supplies one such technique. Roughly speaking, SC allows us to share a considerable portion of the simulation of different designs, and thus achieves a speed up of the total simulation time, no matter if we use a SIMD machine or a regular non-parallel computer.

Philosophically, OO is also different from traditional optimization algorithms which go from point to point. To some extent, this is technologically dictated by older technology when computer memories are expensive and limited. Instead, to a first approximation, memories are costless nowadays and here we can approach the problem basically in parallel. Starting from the entire search space of Θ, we sample uniformly N representative points. Then we select S from N using a crude model with the OO theory guaranteeing the existence of members of G in S. The spirit is a successive narrowing down of the search space. Thus, a parallel implementation of OO is inevitable and natural.

2.1 The concept of the standard clock

The Standard Clock (SC) method (Vakili 1991; Vakili et al. 1992) is an efficient technique for the simulation of parametrically different but structurally similar (PDSS) DEDS. In SC there is no primary sample path. Events for all experiments/simulations are derived from a global event stream. The basic ideas of SC are quite different from traditional approaches, say the Event-Scheduling Simulation (ESS). ESS builds an active event list based on the current state, determines lifetimes for each event in the list, and chooses the event with the minimum lifetime to be the next triggering event for state transition. The cycle repeats with lifetime determination and state transition interacting continuously. By contrast, SC has neither an event list nor event lifetimes. For simplicity, we explain the SC approach by using an M/M/1 queue simulation example with an arrival rate of 0.5 and a service rate of 1.0.

Instead of generating the two types of events (arrival and departure) from separate exponential distributions, we consider a single stream of events that occur at the (faster) rate $0.5 + 1.0 = 1.5$. Namely, the interval time between two events is exponentially distributed with rate 1.5. For each event, a random number $r \in U[0,1)$ is generated to determine the event type. In the following Fig. 7.3, a straight vertical line denotes an event. Each has a $U[0,1)$ random number associated with it.

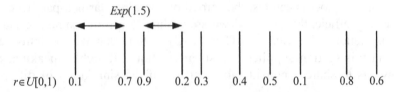

$Exp(1.5)$

$r \in U[0,1)$ 0.1   0.7 0.9   0.2 0.3   0.4   0.5   0.1   0.8   0.6

Fig. 7.3. An example of an event stream before the determination of event types

Because of the properties of Poisson processes, we would expect $0.5/(0.5 + 1.0) = 1/3$ of the events to be arrivals, and 2/3 to be departures. We determine the event type according to the outcome of a $U[0,1)$ random number placed onto a ratio yardstick (Fig. 7.4).

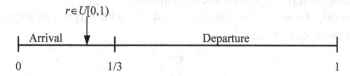

Fig. 7.4. An example of ratio yardstick for determining event types

If $r < 1/3$, this particular event is an arrival event. Otherwise, it could be a departure event. The actual realization of event types of Fig. 7.3 are determined and shown in Fig. 7.5. A down arrow denotes an arrival event and an up arrow denotes a departure event.

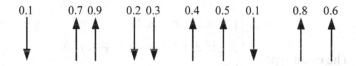

Fig. 7.5. An example of an event stream after the determination of event types

Statistically, this process is equivalent to generating two separate Poisson event streams at rates 0.5 and 1.0. These two event streams, representing the maximal rates of arrival and departure events, are further **thinned** (deleted) according to the state of the DEDS. We ignore departure events whenever the queue is empty, for the events are infeasible. A sample path based on the event stream in Fig. 7.5 is constructed as follows.

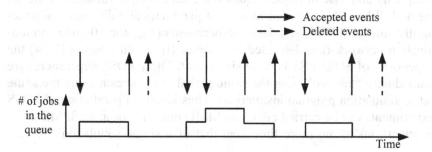

Fig. 7.6. The sample path constructed from the event stream in Fig. 7.5

Because of the memoryless property of the exponential distribution, the sample path constructed in this way is statistically indistinguishable from the path constructed by ESS. The idea of thinning a Poisson event stream can be applied to all networks subject to Markovian assumptions (i.e., all event times must be exponentially distributed).

In general, if we want to simulate a DEDS with n types of exponential events at rates λ_i, $i = 1,..,n$, let

$$\Lambda = \sum_{i=1}^{n} \lambda_i .$$

There are three steps involved:

Step 1. Generate a sequence of events (the event types yet to be determined). The interval time between two events is exponentially distributed with rate Λ.

Step 2. The event type is determined through a $U[0,1)$ random number r for each event as follows:

$$\text{The event type} = \begin{cases} 1 & \text{if } 0 \leq r < \lambda_1 / \Lambda \\ 2 & \text{if } \lambda_1 / \Lambda \leq r < (\lambda_1 + \lambda_2) / \Lambda \\ \vdots & \\ n & \text{if } (\lambda_1 + \lambda_2 + \cdots + \lambda_{n-1}) / \Lambda \leq r < 1. \end{cases}$$

Step 3. Check event feasibility and construct a sample path.

The generation of event streams (Steps 1 and 2) is independent of system states and, therefore, can be done off-line once for all. Given an event stream, we only need to continually check the event feasibility (Step 3) during simulation. This significantly reduces the on-line simulation cost. Simplicity and ease of implementation are additional advantages of the SC method. When SC is applied to a set of parametrically different but structurally similar (PDSS) simulation experiments (e.g., the 10-node communication network first discussed in Chapter III. 3 and Chapter IV. 4) the superiority of SC to ESS is more significant. These PDSS experiments are individually "thinned" from the **same** global event stream using the **same** set of simulation program instructions. Thus ideally, a parallel set of PDSS experiments can be carried out on an SIMD machine, or in an SPMD environment[2], taking no more time than that of a single simulation[3]. A very

---

[2] Stands for Single Instruction Multiple Data and Single Program Multiple Data respectively.
[3] Modulo overhead communication time required in any SIMD machine.

large speed up can be achieved in this way on a massively parallel SIMD machine (Patsis et al. 1997). The metaphor here is the well known trick of data compression and expansion used in the transmission of TV pictures. We can transmit the first frame followed by the "difference" only for succeeding frames. This greatly reduces the transmission load, making moving image reconstruction much more easily implementable than the brute force transmission of two separate frames. Now suppose that we are simulating a sample path denoted by $x(t;\theta,\xi)$ where θ represents the system parameter(s) and ξ all the randomness in the simulation. If we wish to **simultaneously** compute $x(t;\theta+\Delta\theta,\xi)$, i.e., a parametrically different sample path, it is not necessary to repeat all the calculations. A considerable portion of the $x(t;\theta,\xi)$ and $x(t;\theta+\Delta\theta,\xi)$ computation can be shared. Leverage increases with the number of parallel PDSS experiments. This is the essence of the SC approach regardless of whether we are using a massively parallel machine or a regular sequential computer.

In addition, when using a parallel computer, since the SC approach distributes the experiments over multiple processors, the problems of synchronization in the distribution of an inherently sequential simulation algorithm disappear (Fujimoto 1990). Other statistical advantages of common random numbers (Glasserman and Yao 1992), coupling (Glasserman and Vakili 1992), and correlation (Deng et al. 1992) further accrue to such a parametrically different but structurally similar approach of parallel simulation.

2.2 Extension to non-Markov cases using second order approximations

As previously discussed, when event lifetime distributions are exponential, SC is not only an efficient simulation approach, but also easily implementable on computers. This subsection provides efficient approaches for extending the applicability of SC to general distributions (such as uniform distributions), while keeping SC's advantages. The basic idea is as follows. We first argue that it is more important in real world simulation to model the state transition function (i.e., the rules of operation) of the DEDS accurately than to capture the exact distribution of the different event types. Consequently, if we can adequately approximate the first two moments of any event stream and model the rest of the DEDS exactly, the SC approach should qualify as an efficient general purpose simulation language, particularly useful for parallel processing. Based on the algorithm in (Vakili et al. 1992), we develop simple and efficient approximation techniques to accomplish this goal. In Sections 2.2.1, this second order approximation is given. Numerical testing in Section 2.2.2 demonstrates that this second order approach is a good approximation and is faster than parallel ESS.

2.2.1 Second order approximation

In this section, we provide an efficient extension of SC to general distributions using a second order approximation technique. This approximation technique that we are presenting below approximates non-exponential distributions by matching their means and variances and preserves the advantages of the original SC approach. For distributions with means greater than standard deviations, we will describe the use of shifted exponential distributions as approximations. For distributions with means smaller than standard deviations, we will discuss an alternative approach. Note, although the Method of Stages (Kleinrock 1975) can also be applied to approximate non-exponential distributions, it is much less efficient than the shifted exponential ones.

(i) Approximate distributions with means greater than standard deviations

For distributions with means greater than standard deviations, the approach we suggest here is to approximate them by using **shifted exponential distributions**. These can be represented by $K + T$, where K is a constant and T is an exponentially distributed random variable with rate μ (mean = $1/\mu$)[4]. For example, consider an M/G/1 case with non-exponential service time S. We choose K and T such that

$$E[K+T] = E[S],\qquad(7.9)$$

and

$$\mathrm{var}[K+T] = \mathrm{var}[S].\qquad(7.10)$$

From the above two requirements, we have

$$K + \frac{1}{\mu} = E[S],\qquad(7.11)$$

---

[4] If we choose K to be a random variable instead of a constant, we can approximate a non-exponential distribution arbitrarily well such that $K + T$ has the same first n moments as the original random variable by increasing n.

and

$$\frac{1}{\mu^2} = \text{var}[S].$$ (7.12)

From Eq. (7.11) and (7.12), K and μ can be determined as follows.

$$\mu = \frac{1}{\sqrt{\text{var}[S]}},$$ (7.13)

$$K = E[S] - \sqrt{\text{var}[S]}.$$ (7.14)

We generate an event stream with rate $(\lambda + \mu)$ and use a ratio yardstick to determine the arrival and departure events. When a customer's service starts, we set a mask of length K during which all departure events are ignored. In computer simulations, we set a timer when a customer's service begins, and check the timer for each departure event. If the timer is smaller than K, the departure event is ignored. If the timer is greater than K, one exponential event time follows. Since the exponential distribution is memoryless, the following departure event is accepted as a true departure event. Otherwise, this event is ignored. The following figure is given to explain how this approach works.

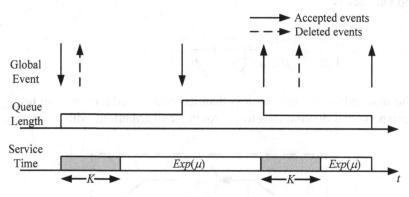

Fig. 7.7. The shifted exponential approach on SC

The second order approach shapes the global event stream by both checking event feasibility, like the original SC approach, and checking the duration of service. The departure event will be ignored (thinned) when either the queue is empty or the duration of service time is smaller than K. This second order approach modifies the event feasibility determination.

Conceptually, we enlarge the system state to be the union of the original state (number of customers in the queue) and the duration of service time. The event feasibility is determined based on the enlarged state. This relaxes the memoryless property of exponential distributions and matches the means and variances of original distributions. Thus we still preserve the primary advantage of the SC approach, that the global event generation is independent of system states. This idea will also be applied to the hyperexponential approach discussed below.

(ii) Approximate distributions with means smaller than standard deviations

For distributions with means smaller than standard deviations, we propose the use of parallel servers (see Fig. 7.8 below) to approximate them. Observe that the density function of a hyperexponential distribution is

$$\alpha \mu_1 e^{-\mu_1 t} + (1-\alpha) \mu_2 e^{-\mu_2 t}, \tag{7.15}$$

with the probability "α" <1. Its mean is

$$\frac{\alpha}{\mu_1} + \frac{1-\alpha}{\mu_2}, \tag{7.16}$$

and variance is

$$2\left(\frac{\alpha}{\mu_1^2} + \frac{1-\alpha}{\mu_2^2}\right) - \left(\frac{\alpha}{\mu_1} + \frac{1-\alpha}{\mu_2}\right)^2. \tag{7.17}$$

The standard deviation is no less than its mean. Therefore, we can use hyperexponential distributions to approximate distributions with larger coef-

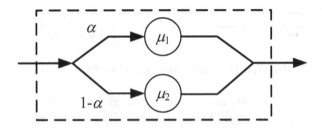

Fig. 7.8. A server with 2 parallel stages

ficients of variation. Since the hyperexponential distribution is equivalent to Fig. 7.8, we see that it can be easily implemented with SC.

For the M/G/1 case, suppose the arrival rate is λ. We select μ_1 and μ_2 to match the mean and variance of the service time. The approximated service time is exponentially distributed either at rate μ_1 or at rate μ_2. We generate three types of events with rates λ, μ_1 and μ_2 respectively. The global clock rate is $\Lambda = \lambda + \mu_1 + \mu_2$. The ratio yardstick for the SC is as follows.

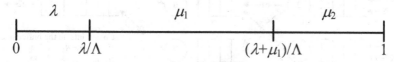

Fig. 7.9. A ratio yardstick for the hyperexponential approach

When a job starts its service, generate a $U[0,1)$ random number u. If $u < \alpha$, the job is served at rate μ_1, so we accept the μ_1 event as a departure event, with the μ_2 event ignored. On the other hand, if $u \geq \alpha$, the job is served at rate μ_2, and we accept the μ_2 event as a departure event, with the μ_1 event ignored. A more efficient way for the hyperexponential approach is to modify the ratio yardstick as follows. Without loss of generality, assume μ_1 is greater than μ_2.

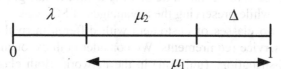

Fig. 7.10. A more efficient ratio yardstick for the hyperexponential approach

Note that the global clock rate is reduced to $\Lambda = \lambda + \mu_1$, although there are three types of events: λ, μ_2, and Δ. The procedures of event type determination are also modified.

Unlike the shifted exponential approach, the hyperexponential approach shapes the global event stream by checking both the event feasibility and the service type (rate μ_1 or μ_2). This is another example of the general idea of using "state" information (in this case number of customers and event type) to thin the global event stream.

2.2.2 Numerical testing

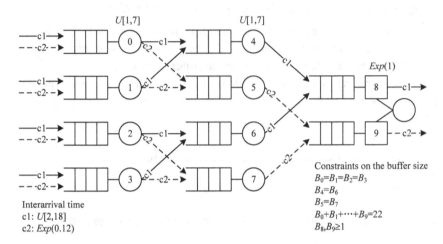

Fig. 7.11. A 10-node network with priority and shared server

Consider the 10-node network shown in Fig. 7.11. This is an example we submitted as a prize winning entry to the 1992 MasPar Challenge. For details about this example, please refer to (Patsis et al. 1997). We have used this example in Chapter III Section 3 and Chapter IV Section 4 to demonstrate the selection rules and Vector Ordinal Optimization (VOO). Here our purpose is to demonstrate that our approach has good approximation properties while preserving the advantages of SC.

There are two classes of customers with different arrival distributions but the same service requirements. We consider both exponential and non-exponential distributions (uniform) in the network. Both classes arrive at any of the 0-3 nodes, and leave the network after having gone through three different stages of service. The routing is not probabilistic, but class dependent as shown in Fig. 7.11. Finite buffer sizes at all nodes are assumed, which is exactly what makes our optimization problem interesting. Namely, we are interested in distributing optimally buffer spaces to different nodes given a limited budget for them. A buffer is said to be full if there are as many customers as its size in it, not including the customer being served in the server. Blocking in the network is governed by the following rules:

(i)   If a buffer at any of the entry points (where customers from outside enter, that is nodes 0-3) is full, no more arrivals are allowed at the corresponding node.

(ii) If a buffer at any of the inside nodes (nodes 4-9) is full, the customer(s) that are about to enter this buffer are prevented from departing the server where they just received service. It results in keeping the aforementioned server idle until the customer finally finds an empty slot in the buffer downstream and departs.

(iii) If more than one server is blocked by the same buffer downstream (i.e., buffer 9 blocking both servers 5 and 7), First Blocked First Served is applied.

More specifically, class 1 customers' interarrival periods are distributed uniformly from 2 to 18; and class 2 customers' interarrivals are distributed exponentially with rate 0.12. The service time for both classes is uniform from 1 to 7 at nodes 0-7 (two first stages of service) and exponential with rate 1.0 at the last stage (nodes 8-9). Nodes 8 and 9 each have their own queues, but they share a single server governed by the following rules:

(iv) If the length of the queue at node 8 is greater than one, a node 8 customer is served. If a customer at node 9 is being served and the length of node 8's queue becomes greater than one, service is interrupted to allow the higher priority, class 1 customer to be served. Otherwise, customers are served on a first-come first-served basis.

Furthermore, we include a priority system for the two classes. At buffers 0-3, class 1 customers now jump ahead of any class 2 customers in the queue. If a class 2 customer has already begun service and a class 1 customer arrives, the class 2 customer is allowed to complete service.

Such a network could be the model for a large number of real-world systems, such as a manufacturing system, and a communication or a traffic network. We consider the problem of allocating 22 buffer units, among the 10 different nodes numbered from 0 to 9. We denote the buffer size of node i as B_i. We set constraints as shown in Fig. 7.11 for symmetry reasons. Totally, there are 1001 different configurations. We want to analyze the throughputs of them.

A simulation is run by both the ESS and SC. The system throughput and average system time are computed for comparison. Let P_i and $\hat{P}_i$ be the system throughputs of design i estimated by ESS and SC; T_i and $\hat{T}_i$ the average system time estimated by ESS and SC, respectively. The average error of estimated throughput is

$$\frac{1}{1001} \sum_{i=1}^{1001} \frac{\hat{P}_i - P_i}{P_i}$$

and the average error of estimated system time is

$$\frac{1}{1001} \sum_{i=1}^{1001} \frac{\hat{T}_i - T_i}{T_i} .$$

The simulation is stopped after 250,000 customers have left the system. From the simulation results, we find that the average error of estimated throughput is 1.4% and the average error of estimated system time is 0.12%. They show that the 2nd order approximation of SC has good approximation properties.

On the other hand, the following experiments demonstrate that SC is faster than ESS. The above problem is simulated in parallel on a MasPar-1 SIMD machine, in which there are 1024 processors. The CPU times for different lengths of simulation are listed in Table 7.2. SC is more than three times faster than the traditional approach, which illustrates the saving due to sharing of computation under the SC approach.

Table 7.2. Parallel event-scheduling vs. parallel SC (Chen and Ho 1995) © 1995 IEEE

# of Customers that left the system	Event-scheduling CPU Time (in sec)	SC Simulation CPU Time (in sec)
100	5.48	1.37
1000	53.80	15.01
10000	536.34	152.60
50000	2682.05	761.12
100000	5363.59	1522.37
150000	8045.95	2283.51
200000	10728.31	3044.52
250000	13409.84	3802.14

3 Effect of correlated observation noises

One of the assumptions of both the analytical and experimental results of ordinal optimization we discussed so far is that the performance estimation errors (or the observation noises) are independently and identically distributed (i.i.d.). In practical applications such as performance estimation for DEDS via simulation, this assumption of independent estimation error from one design to next may not hold due to the use of common random variables, replications with identical initial conditions, and parallel simulation (Vakili et al. 1992), etc. Naturally there arises one question: Can we still apply the UAP table in Section II.5 to determine the size of the selected set when the observation noises are correlated with each other, or when the noises are not identically distributed and θ dependent? This is the question we are going to answer in this section. The quick answer is yes. Actually we will show that the correlation among the observation noises seldom hurts but usually helps. Practitioners can now skip the rest of this section and apply OO without worrying about the correlations among observation noises. For readers interested in details, we show how the correlation helps as follows.

First, consider the extreme case of positive correlation. In other words, all the observation noises are perfectly correlated, i.e., they are identical. The observed order is always the true order. The observed good enough designs are all truly good enough. The correlation helps a lot in this case. Second, consider the extreme case of negative correlation. Since in ordinal optimization we care about the observed order among the designs more than the observed value, the worst case is that the observation noises of adjacent designs (w.r.t. true performances) are perfectly but negatively correlated. The designs can be separated into two halves. Within each half, the observation noises are perfectly correlated. This is like the removal of half the designs from consideration. At least half of the observed top-n% designs are truly top-n% of the entire design space. The correlation helps in this case, too.

In the general case, when the observation noises of some designs are positively correlated, with others negatively correlated, the analysis becomes complicated. To get a rough idea, we consider the simple case of only two designs first. Suppose there are two designs θ_1 and θ_2. The true performances are $J(\theta_1)$ and $J(\theta_2)$, respectively. The observation noise is w_i, i.e.,

$$\hat{J}(\theta_i) = J(\theta_i) + w_i, \text{ for } i = 1 \text{ and } 2.$$

Suppose vector $[w_1, w_2]^\tau$ contains two-dimensional normal distribution with mean and covariance matrix as follows

$$\begin{bmatrix} 0 \\ 0 \end{bmatrix} \text{ and } \begin{bmatrix} \sigma_{11} & \sigma_{12} \\ \sigma_{12} & \sigma_{22} \end{bmatrix}.$$

Suppose design θ_1 is better, i.e., $J(\theta_1) < J(\theta_2)$, and we want to predict which design is better based on the observed performance. Then the probability that we make correct prediction is

$$\text{Prob}\left[\hat{J}(\theta_1) < \hat{J}(\theta_2)\right]$$
$$= \text{Prob}\left[J(\theta_1) + w_1 < J(\theta_2) + w_2\right]$$
$$= \text{Prob}\left[w_1 - w_2 < J(\theta_2) - J(\theta_1)\right]$$
$$= \frac{1}{\sqrt{2\pi\left(\sigma_{11} + \sigma_{22} - 2\sigma_{12}\right)}} \int_{-\infty}^{J(\theta_2) - J(\theta_1)} \exp\left(-\frac{x^2}{2\left(\sigma_{11} + \sigma_{22} - 2\sigma_{12}\right)}\right) dx$$
$$= \frac{1}{\sqrt{2\pi}} \int_{-\infty}^{(J(\theta_2) - J(\theta_1))/\sqrt{\sigma_{11} + \sigma_{22} - 2\sigma_{12}}} \exp\left(-\frac{x^2}{2}\right) dx.$$

It is clear that this probability increases when σ_{12} ($=\sigma_{21}$) increases. Roughly speaking, this means the more w_1 and w_2 are correlated with each other, the more likely we pick up the right design θ_1.

When there are a lot of designs, the analysis becomes very complicated and tedious, if not impossible. Hence we use experiments to observe the effect of the correlated observation noises on the alignment between the observed good enough designs and truly good enough designs. Suppose there are $N = 200$ designs, with true performances within $[0, 200]$. The observation noises w's are linear combinations of i.i.d. normally distributed noises, $v_1, v_2, \ldots v_N$, i.e.,

$$\begin{pmatrix} w_1 \\ \vdots \\ w_N \end{pmatrix} = A \begin{pmatrix} v_1 \\ \vdots \\ v_N \end{pmatrix}, \tag{7.18}$$

where A is an $N \times N$ matrix. The normal distribution assumption can be easily justified based on the central limit theorem for most cases. Assume the noise is relatively large, compared with the largest performance value 200, and let $v_1, v_2, \ldots v_N \sim N(0, 10000/12)$. Suppose $w_1 = v_1$, $w_{i+1} = aw_i + bv_{i+1}$ ($i = 1, \ldots, N-1$), and a and b are two constants such that $a^2 + b^2 = 1$. Then the variances of all w_i's are the same. Matrix A in Eq. (7.18) is then given by

$$A = \begin{pmatrix} 1 & 0 & 0 & \cdots & 0 \\ a & b & 0 & \cdots & 0 \\ a^2 & ab & b & \cdots & 0 \\ \vdots & \vdots & \vdots & \cdots & \vdots \\ a^{N-1} & a^{N-2}b & a^{N-3}b & \cdots & b \end{pmatrix}.$$

We can easily verify that the correlation among the observation noises is increasing with respect to *a*. To cover a wide range of optimization problems, we consider a generic set of the Ordered Performance Curve parameterized by β according to the following

$$J(\theta_i) = \begin{cases} \dfrac{\beta C}{(1-\beta)N}i & i < (1-\beta)N \\ \dfrac{(1-\beta)C}{\beta N}i + \dfrac{(2\beta-1)C}{\beta} & i \geq (1-\beta)N. \end{cases} \tag{7.19}$$

where $C = 200$. The OPCs are illustrated in Fig. 7.12. Fig. 7.12 shows that the OPC changes from the flat type to the steep type, when β increases from 0 to 1.

We vary the values of β and *a* as follows:

$\beta = 0.1, 0.2, 0.3, 0.4, 0.5, 0.6, 0.7, 0.8, 0.9.$

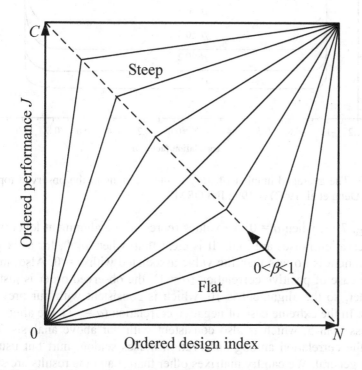

Fig. 7.12. A generic set of ordered performance curves (Deng et al. 1992) © 1992 INFORMS

$a = 0, \pm0.1, \pm0.2, \pm0.3, \pm0.4, \pm0.5, \pm0.6, \pm0.7, \pm0.8, \pm0.9, \pm0.91, \pm0.92,$
$\pm0.93, \pm0.94, \pm0.95, \pm0.96, \pm0.97, \pm0.98, \pm0.99, \pm1.$

For each parameter setting, we show the number of truly top-10 designs contained in the observed top-10 designs in Fig. 7.13, which has averaged over 1100 experiments.

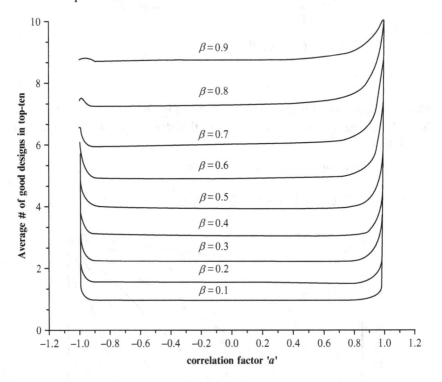

Fig. 7.13. The averaged number of truly top-ten designs in the observed top-ten designs (Deng et al. 1992) © 1992 INFORMS

In Fig. 7.13, when $a = 0$, the values represent the alignment level when the observation noises are i.i.d.. It is clear that, when $|a|>0$, for a fixed β, the alignment is no less than the value associated with $a = 0$. Also, in the extreme case of positive correlation ($a = 1$), the observed order is just the true order, so the alignment is 10, which is consistent with our previous analysis. In the extreme case of negative correlation ($a = -1$), the alignment is no less than 5, which is also consistent with our above analysis. This means the correlation among observation noises seldom hurt but usually helps in general. We can try matrixes other than A and the results are similar (Deng et al. 1992).

In conclusion, we demonstrated in this section that the correlations among the observation noises seldom hurt but usually helps. We in general can use the UAP table in Section II.5 to predict the size of the selected set, and with little worry in practice.

Before the end of this section, we add a short discussion on another case of non-i.i.d. noise, which is independent non-identical observation noise. Astute readers may ask whether this kind of noises causes us any trouble in the application of OO in practice. The answer is no. To see this, just notice that, if we choose the noise level with respect to the noise with the largest variance, then we can calculate the size of the selected set using the UAP table in Section II.5. This size is an upper bound of the size that is actually needed, because the observation noises of some designs are not that large. A more interesting question is: Can we take advantage of the independent non-identical noise? The answer is yes. The basic idea is to divide the design space into several sub-regions. By estimating the average performance of the designs within each region, we combine this information into the crude model, and then hopefully the observation noise w.r.t. the new crude model is (closer to) i.i.d., with a smaller noise level. This yields a smaller set S to select. Of course, it takes computing budget to estimate the average performance of the designs within a region. Thus, either spending more computing budget to get a better crude model, or spending the computing budget to deal with a larger selected set, there is a trade off. For more details, please refer to (Yang et al. 1997), (Yang 1998) and (Yang and Lee 2002). Of course, in general, we may have non identical and correlated noises. We can always use the maximum noise level to evoke the conclusion in this section. Together with the previous paragraph, we now can ensure practitioners to apply OO with little worry about the i.i.d. assumption of the observation noises.

4 Optimal Computing Budget Allocation and Nested Partition

In this section we discuss two algorithms that are coherently related to OO, the optimal computing budget allocation (OCBA) which serves as a specific selection rule of OO, and nested partition (NP) which serves as a possible framework to do OO iteratively. Besides these two algorithms, OO can also be combined with many others, which will be further discussed in Section 6.

4.1 OCBA

As we argued previously, it is time consuming to evaluate the performance of a design precisely through Monte Carlo simulation. Since the total computing budgets available for performance evaluation (measured by the total number of replications) of all designs under consideration are limited, it is natural to ask the question: how can we allocate the computing budgets to individual designs? OCBA was developed by Chen et al. for further enhancing the efficiency of OO (Chen et al. 2000). The goal is to utilize optimally the given computing budgets to achieve a high alignment probability. In other words, we want to control the allocation of the computing budgets to all the designs so that the alignment probability is maximized. In Section II.5, we have shown that it is very difficult (if not impossible) to obtain the closed-form expression for the alignment probability, when the selection rule is not Blind Pick. OCBA considers the cases when the good enough design set is defined as the singleton, the true optimum, then the alignment probability is simplified to the probability that the observed best design is the truly best, which is known as the probability of correct selection and denoted as Prob[CS]. There exists a large literature on assessing Prob[CS] based on classical statistical models (e.g., (Goldsman and Nelson 1994) and (Banks 1998) give an excellent survey on available approaches). However, most of these approaches are only suitable for problems with a small number of designs. For the applications that OO is used, there is usually a large design space, say 10^8 or at least of the size $N = 1000$ if uniform sampling in design space is employed. The computing budgets of those approaches soon become practically infeasible. For these reasons, OCBA was developed through a series of development.

Suppose there are N designs $\theta_1, \ldots \theta_N$ with true performances $J(\theta_1), \ldots J(\theta_N)$. The observed performance is

$$\hat{J}(\theta_i) = J(\theta_i) + w(\theta_i),\tag{7.20}$$

where for each design $w(\theta_i)$ contains i.i.d. normal distribution $N(0, \sigma_i^2)$. Suppose we can run simulation no more than T times during the selection process of OO. By the time when the selection process of OO ends, design θ_i is observed n_i times, and $\sum_{i=1}^{N} n_i = T$. If we use the mean value of the observations to evaluate the performance of a design, we have

$$\overline{\hat{J}}(\theta_i, n_i) = J(\theta_i) + w(\theta_i, n_i),\tag{7.21}$$

where $w(\theta_i, n_i)$ contains normal distribution $N(0, \sigma_i^2/n_i)$. The problem that OCBA tries to solve is

$$\max_{n_1, n_2, \ldots n_N} \text{Prob}[\text{CS}] = \text{Prob}\left[\bar{\hat{J}}(\theta_b, n_b) < \bar{\hat{J}}(\theta_i, n_i), \text{for all } i \neq b\right]$$

$$\text{s.t. } \sum_{i=1}^{N} n_i = T,$$

(7.22)

where θ_b is the truly best design. The difficulty to solve the above maximization problem lies in the calculation of Prob[CS]. To avoid this difficulty, a lower bound of Prob[CS] is used as the objective function to be maximized. When the lower bound is maximized, we hope that the resultant Prob[CS] will not be too bad. To construct the lower bound, note that, for a set of random variables Y_i, it holds that

$$\text{Prob}\left[\bigcap_{i=1}^{N}(Y_i < 0)\right] = 1 - \text{Prob}[Y_1 < 0 \text{ or } Y_2 < 0 \text{ or } \ldots Y_N < 0]$$

$$= 1 - \left(\text{Prob}[Y_1 < 0] + \text{Prob}[Y_1 \geq 0, Y_2 < 0] + \ldots\right.$$

$$\left. + \text{Prob}[Y_1 \geq 0, \ldots Y_{N-1} \geq 0, Y_N < 0]\right)$$

$$\geq 1 - \left(\text{Prob}[Y_1 < 0] + \text{Prob}[Y_2 < 0] + \ldots + \text{Prob}[Y_N < 0]\right)$$

$$= 1 - \sum_{i=1}^{N}\left(1 - \text{Prob}[Y_i \geq 0]\right).$$

This is known as the Bonferroni inequality. In our case, $Y_i = \bar{\hat{J}}(\theta_b, n_b) - \bar{\hat{J}}(\theta_i, n_i)$. Then

$$\text{Prob}[\text{CS}] \geq 1 - \sum_{i=1}^{N}\left(1 - \text{Prob}\left[\bar{\hat{J}}(\theta_b, n_b) - \bar{\hat{J}}(\theta_i, n_i) \geq 0, i \neq b\right]\right), \quad (7.23)$$

which is referred to as the approximated probability of correct selection (APCS). So the problem in Eq. (7.22) is converted to

$$\max_{n_1, n_2, \ldots n_N} 1 - \sum_{i=1}^{N}\left(1 - \text{Prob}\left[\bar{\hat{J}}(\theta_b, n_b) - \bar{\hat{J}}(\theta_i, n_i) \geq 0, i \neq b\right]\right)$$

$$\text{s.t. } \sum_{i=1}^{N} n_i = T.$$

(7.24)

This is the problem solved by OCBA. (Chen et al. 2000) has shown that APCS can be asymptotically maximized when $T \to \infty$ if

$$\frac{n_i}{n_j} = \left(\frac{\sigma_i / \delta_{b,i}}{\sigma_j / \delta_{b,j}} \right)^2 , i, j \in \{1, 2, ..., N\}, \text{ and } i \neq j \neq b \qquad (7.25)$$

$$n_b = \sigma_b \sqrt{\sum_{i=1,i\neq b}^{N} \frac{n_i^2}{\sigma_i^2}} \qquad (7.26)$$

where $\delta_{b,i} = J(\theta_b) - J(\theta_i)$. Of course we do not know the true performances $J(\theta_i)$ during the optimization, so we use the mean value of the observations as an estimate. Intuitively, in Eq. (7.25), if a design is estimated to be bad, which means $\delta_{b,i}$ is large, n_i should be small, which means we should not allocate more computing budgets to this bad design. If σ_i is large, which means the observation noise is still large, we should allocate more computing budget to this design to obtain better performance estimate. There is a tradeoff, which is what the term $\sigma_i / \delta_{b,i}$ represents. We can also show that Eq. (7.25) and Eq. (7.26) are intuitively reasonable for simple cases. Say there are only two designs θ_1 and θ_2, and $J(\theta_1) < J(\theta_2)$, i.e., $b = 1$. Then based on Eq. (7.26) we have $n_1 = \sigma_1 \sqrt{n_2^2 / \sigma_2^2}$. Therefore, $n_1/n_2 = \sigma_1/\sigma_2$. In what follows, we can easily testify that this allocation is identical to the optimal allocation solution when there are only two designs. Suppose there are only two designs, and $J(\theta_1) < J(\theta_2)$, then

$$\text{Prob}[\text{CS}] = \text{Prob}\left[\bar{\bar{J}}(\theta_1, n_1) < \bar{\bar{J}}(\theta_2, n_2) \right]$$
$$= \text{Prob}\left[w(\theta_1, n_1) - w(\theta_2, n_2) < J(\theta_2) - J(\theta_1) \right],$$

where $w(\theta_1, n_1) - w(\theta_2, n_2) \sim N(0, \sigma_1^2/n_1 + \sigma_2^2/n_2)$. To maximize Prob[CS], we need to minimize $\sigma_1^2/n_1 + \sigma_2^2/n_2$ subject to the constraint that $n_1 + n_2 = T$. Solving this optimization problem, we have $n_1/n_2 = \sigma_1/\sigma_2$.

The allocation in Eqs. (7.25) and (7.26) are asymptotically optimal, and the estimate of $\delta_{b,i}$ improves when more observations are taken. We can do the allocation in an iterative way. Each time we only decide how to allocate a small amount of our computing budgets, and update the estimate of $\delta_{b,i}$ and σ_i successively. The procedure of OCBA is shown in Box 7.1.

Box 7.1. Procedures of OCBA

> Step 1: Perform n_0 simulation replications for all designs. $l = 0$, $n_1^l = n_2^l = ... = n_N^l = n_0$.
>
> Step 2: If $\sum_{i=1}^{N} n_i^l \geq T$, stop.
>
> Step 3: Increase the computing budget by Δ and compute the new budget allocation, $n_1^{l+1}, n_2^{l+1}, ..., n_N^{l+1}$, using Eqs. (7.25) and (7.26)
>
> Step 4: Perform additional $\max\left(0, n_i^{l+1} - n_i^l\right)$ simulations for design i, $i=1,...,N$, $l+1 \rightarrow l$. Go to step 2.

If we equally allocate the computing budgets to all the designs, and select the observed best design as an estimate of the truly best, this is a special case of the horse race rule (HR). To show how OCBA helps to enhance the performance of OO, we compare OCBA and HR through a series of experiments, including:

Problem 1: Normal distribution. Suppose there are 10 designs with true performances $J(\theta_i) = i$, and i.i.d. observation noise $w(\theta_i) \sim N(0, 6^2)$, $i = 0, 1, ...9$.

Problem 2: Uniform distribution. Suppose there are 10 designs with true performances $J(\theta_i) = i$, and observation noise $w(\theta_i) \sim U(-10.5, 10.5)$, $i = 0, 1, ...9$.

Problem 3: Normal distribution with larger variance. Suppose there are 10 designs with true performances $J(\theta_i) = i$ and i.i.d. observation noise $w(\theta_i) \sim N(0, 2 \times 6^2)$, $i = 0, 1, ...9$.

Problem 4: Flat OPC. Suppose there are 10 designs with true performances $J\left(\theta_i\right) = 9 - 3\sqrt{9-i}$, and i.i.d. observation noise $w(\theta_i) \sim N(0, 6^2)$, $i = 0, 1, ...9$.

Problem 5: Steep OPC. Suppose there are 10 designs with true performances $J\left(\theta_i\right) = 9 - \left(\dfrac{9-i}{3}\right)^2$, and i.i.d. observation noise $w(\theta_i) \sim N(0, 6^2)$, $i = 0, 1, ...9$.

Problem 6: Bigger design space. Suppose there are 100 designs with true performances $J(\theta_i) = i/10$, and i.i.d. observation noise $w(\theta_i) \sim N(0, 1^2)$, $i = 0, 1, ...99$.

In OCBA we set $n_0 = 10$ and $\Delta = 20$. In HR we allocate the computing budget T equally to all the designs. We incrementally increase the value of T, record the minimal value of T s.t. the Prob[CS]$\geq$0.99 for OCBA and HR in the above problems, and show the results in Table 7.3.

Table 7.3. Computing burden for Prob[CS]>0.99 by OCBA and HR in the experiments

Problem index	OCBA	HR	Speed up ratio = HR/OCBA
Normal distribution	1100	4400	4.0
Uniform distribution	1900	6000	3.2
Normal distribution with larger variance	2100	8500	4.0
Flat OPC	4100	15100	3.7
Steep OPC	300	1100	3.7
Bigger design space	2600	39000	15

From Table 7.3, we can see that OCBA saves the computing budgets and thus speeds up the optimization at least 3 times in all the problems tested, even if the noise distribution is not normal (Problem 2). Especially when the design space is larger (Problem 6), OCBA achieves a larger speed up. This is because a larger design space gives the OCBA algorithm more flexibility in allocating the computing budgets. This comparison shows that OCBA can further enhance the performance of OO.

OCBA has been successfully applied to many problems, such as the communication network (Chen 1994), the reliability optimization of the transportation system capacity (Lin 2004), the manufacturing scheduling problems (Chen et al. 1997), Monte Carlo simulation-based product design (Chen et al. 2003), the walking robot motion planning problem (Luo 2000), the buffer allocation problem (Shi and Chen 2000), and the combi- nation with rank and selection (Chen 2004), just to name a few. Interested readers please refer to our online reference list (Shen and Bai 2005) and specific papers for more technical details.

4.2 NP

Nested partition, as the name shows, is a way to successively narrow down the search space (Ólafsson and Shi 1999; Shi and Ólafsson 2000a, 2000b). NP was developed to deal with the difficulty of large design space that was present in many combinatorial problems. OO can also be regarded as a way to narrow down the search space (first from the large design space to the N designs that are sampled as a representative of the design space, then from the N designs to the selected set). Since we have only applied OO once in previous chapters, it is natural to ask how we should apply OO iteratively. Though this is still an open question, there are some ideas (Deng and Ho 1997). The combination of NP and OO supplies a possible way to do OO iteratively.

The idea of NP is based on the evolution of the division of the entire design space into two regions, namely the promising region and the surrounding region. Ideally the promising region should *only* contain the designs we are searching for. The task of NP is to adjust the division between the two regions to achieve this. A measurement based on observations of a small subset of designs in a region is used as a promising index to guide the adjustment of division between promising and surrounding regions[5]. Backtracking is allowed to avoid the division stuck at local optima.

More specifically, in each iteration, NP contains four major steps. **First**, partitioning. When the entire design space is divided into several disjointed subregions, only one subregion is indicated as the promising region. In the beginning, the entire design space is not divided and is the promising region. In the partition step, the promising region is partitioned into several disjointed subregions, and the other subregions surrounding the promising region (if any) are aggregated into one region, called the *surrounding region*. **Second**, sampling. In each of the subregions (including the ones partitioned from the promising region and the surrounding region), we randomly sample some designs. This step is rather flexible. The only requirement is that each point in the design space should have some positive probability to be sampled. This will avoid getting stuck in a local minimum. **Third**, identifying the new promising region. Based on the observed performances of the designs sampled from each subregion, we identify one region as the new promising region for further partition. Many rules can be used in this step. One idea is to select the subregion that contains some observed good enough designs. If there are more than one subregions containing observed good enough designs, the tie is broken uniformly and randomly. If the surrounding region is identified as the promising region, we need a **fourth** step: backtracking. There are different rules to do backtracking. The idea is also to avoid getting stuck in the local minimum. One rule is to move back along the partition history by one iteration, i.e., to revoke the partition used in this iteration and return to the previous partition situations. Another rule is to move back to the entire design space, then there is only one region. After the four steps, we obtain a new promising

---

[5] The implied structural assumption here is that, if a region has many good designs, it probably contains even more good or better designs – a kind of continuity of neighborhood assumption.

region and goes back to step 1 for further partitioning[6]. This process continues until all the computing budgets are consumed, or the promising region is a singleton and there is no further partition (Box 7.2).

Box 7.2. NP procedures

Step 1: Initialization. The entire design space is the promising region.
Step 2: Partitioning. If the promising region is not singleton, partition the region into several subregions, and aggregate the surrounding region (if any) into one region.
Step 3: Sampling. Sample designs from each subregion and obtain the observed performances of all the sampled designs. Computing budget allocation procedures may be applied.
Step 4: Identifying the new promising region. If the new promising region is the surrounding region, just backtrack.
Step 5: Going to step 2, unless the new promising region is a singleton, or all the computing budgets are consumed. If the new promising region is a singleton, output that design. Otherwise, output the design that is most frequently visited.

In the above procedure, ideally each time we are able to select the subregion that contains the global optimum. The promising region converges to a singleton very soon.

It is clear that the way to divide the entire design space into subregions is critical to the efficiency of NP. A subregion containing the truly best design is NP-friendly if it provides a direct promising index to indicate the region is good. A subregion is not friendly if its promising index is misleading, i.e., indicates that this region is worse than the surrounding region (complementary set) which does not contain the best design. To quantify the quality of a way to generate a subregion and its surrounding region, we measure the probability that the best of uniformly sampled designs from the subregion is not better than the designs chosen from the surrounding region. Its lower bound is

$$1 - r^n$$

where n is the number of samples blindly picked from the subregion of interest, and r, known as the "overlap" in NP terminology, is the fraction of designs that are truly worse than the designs outside that region. For example, a subregion with 90% overlap means only 10% of the designs in

---

[6] There is another implied assumption in NP, that is, it is relatively easy to characterize the partitioned region via simple inequalities. In other words, the repeated partitioning and aggregation can be easily carried out.

this subregion are truly better than all the designs outside this subregion. Of course, the larger this overlap is, the more difficult it is to identify the promising region correctly as shown in Fig. 7.14.

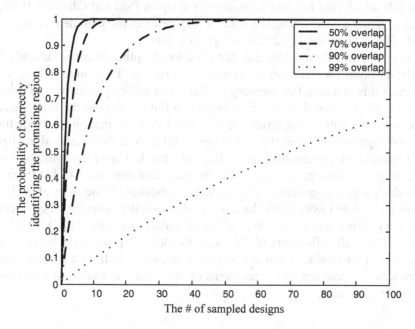

Fig. 7.14. Probability of correctly identifying the promising region converges exponentially fast w.r.t. the number of sampled designs (assuming uniformly sample designs)

In Fig. 7.14, the *x*-axis represents the number *n* of designs sampled from the subregion that contains the truly optimum. The *y*-axis represents the probability that the promising region identified by the NP is the truly promising region. It is important to observe that all the curves in Fig. 7.14 converge to 1 exponentially fast, although for higher level of overlap, the convergence is relatively slow. In the above analysis, we assume that the true performance of each design is available for samples. When observation contains noises, we can still have similar conclusion thanks to the exponential convergence in comparison using order information.

NP can be used to solve both the deterministic optimization problem (say the Traveling Salesman problem (Shi and Ólafsson 2000a)) and the simulation-based optimization problem. When there are observation noises, the selection rules such as OCBA can be used to allocate the computing budgets among the designs so that the observed best design has a larger chance to be the truly best design in that region (Shi and Chen 2000). There are many successful applications of NP, such as in the Traveling Salesman problem

(Shi and Ólafsson 2000a), the discrete resource allocation in supply chain management (Shi et al. 1999), the scheduling in shop floor control (Yoo et al. 2004), combinations with ranking-and-selection (Ólafsson 1999), and the job scheduling for parallel computer systems (Shi and Ólafsson 1998), just to name a few. Interested readers may refer to the complete reference list of OO (Shen and Bai 2005) for specific papers.

It is worthwhile to point out that traditional optimization is basically a point-to-point (design-to-design) iterative process. To a large extent, we suspect this is a result of memory limitation in earlier generations of computers. At any one time, we can only keep limited data in the high speed memory of a CPU. As a result, we only use local information to decide the next design to explore in the search space. Of course, optimizers all realize the importance of past data in guiding the search. Under the limitation of memory, we attempt to summarize the past data into rules to help guide our design-to-design search. One prominent example is the rules of TABU search (Glover 1989). With the advent of essentially unlimited high speed memory, however, we are free of these earlier constraints. OO, OCBA, and NP are all reflections of this new freedom. Optimization becomes a process of narrowing down the search. Thus, we now have a complementary way of approaching the problems of optimization. Many exciting possibilities lie ahead.

5 Performance order vs. performance value

Comparing with the results in the rest of the book, this section and the next Section 6 contain mainly preliminary results instead of complete analysis. We discuss two questions in these two sections: how can we apply OO if we want to find some designs with true performance value close to the true optimum? How can we combine OO with other optimization algorithms? There are no general answers to either question, which is still open to discussion and further research. We will just report some results so far obtained, which are mainly preliminary and for specific problems. They indicate what are possible rather than to provide any definitive extension of the OO methodology.

In this section we report some work in (Lin 2000a)[7]. In conventional ordinal optimization, as introduced in Chapter II, the good enough set is defined by the user as the truly top-g% designs. In other words, the definition

---

[7] We did some modifications though. For example, (Lin 2000a) considers a maximization problem, and we modified the problem formulation to a minimization problem here. But the basic ideas in (Lin 2000a) are preserved.

is ordinal. (We thus denote the good enough set as G_o.) However, in some practical applications, we want to find designs not too much worse in value than the global optimum. For example, we know $\min_{\theta \in \Theta} J(\theta) = 0$, and $\max_{\theta \in \Theta} J(\theta) = 1$, and want to find a design with true performance no greater than 0.1.We should define the good enough set as the designs within top-$g\%$ true performances (say $g\% = 10\%$), i.e., the good enough set is defined according to the cardinal values, instead of order. (So we denote the good enough set as G_v.) Natural questions then arise: Can we apply OO on this type of problems? If so, how can we do that? The answers are not complete yet. However, we have some ideas on how to apply OO when the definition of good enough is based on performance "value" instead of "order", which will be discussed in the rest of this section. A quick answer is a qualified yes. We can apply OO on this type of problems. There are at least two methods for such an application: either to define an appropriate G_o based on the information of G_v and convert the problem to the one that we have already solved in the previous chapters (i.e., the idea of problem conversion), or as introduced in (Lin 2000a), to develop another way to calculate the alignment probability. In both ways, more problem and structural information are required. We introduce these two ways in the following paragraphs and have a special focus on the second one.

First, let us look at the idea of problem conversion. The key is to estimate the size of G_v, $|G_v|$. We can define the good enough set as the designs within top-$|G_v|/|\Theta|$ w.r.t. true performance **order**. In this way we obtain a definition of good enough set w.r.t. order, and convert the problem into the type that we are familiar and has been discussed throughout the previous chapters. The rest of the application of OO is then straightforward. The question is how we can estimate $|G_v|$? This might be difficult, especially when the observation noises are large. One possible way is to uniformly and randomly sample N designs from the entire design space, use the crude model to estimate quickly the performances of these designs, and obtain an observed ordered performance curve. We can then calculate the number of designs within top-$g\%$ observed performances, and use this number divided by N as an estimate of the value of $|G_v|/|\Theta|$. Obviously, this estimate of $|G_v|/|\Theta|$ may not be accurate, but we can use this in practical applications as a rough idea and choose a relatively conservative definition of G_o. In other words, the weakness of this idea is the accuracy of the estimate of the size of G_v. Problem information will help us to improve the accuracy of the estimate.

In the rest of this section, we focus on the second way: developing a new way to calculate the alignment probability. Since the UAP table in Section II.5 is used to estimate the selected size when the good enough set is defined in performance order not value, we need to find another way to calculate the alignment probability and thus the selected size in this new problem. Suppose we know the probability that a selection rule successfully picks a design in G_v, denoted as $P_{success}$. After n independent trials, the selection rule will pick out n designs. The probability that at least one of these n designs is truly good enough in value (i.e., in the set G_v) is $1-(1-P_{success})^n$, which converges to 1 exponentially fast w.r.t. the number of trials n. If we want this probability to be no less than a given value P_0, the minimal n will be

$$n = \left\lceil \frac{\ln(1-P_0)}{\ln(1-P_{success})} \right\rceil \tag{7.27}$$

where $\lceil \bullet \rceil$ is the ceiling function.

Exercise 7.2: Proof this.

For example, if $P_0 = 0.95$, $P_{success} = 0.1$, the minimal n is 29, which means if we use a selection rule with $P_{success} = 0.1$ to pick a design each time, and repeat this process 29 times, then at least one of these 29 selected designs will be truly good enough in performance value with probability no less than 0.95. This result is quite similar to the one we have achieved in the previous chapters when the good enough set is defined in order. The key problem now is how we can estimate $P_{success}$. Though the answer is problem dependent, we will show how to obtain the lower bound of $P_{success}$ in various examples. The first example considers a wide range of optimization problems with different OPCs. It is shown that $P_{success}$ is affected by the selection rule thus used. Under some assumption on the selection rule, we obtain a lower bound of $P_{success}$ that is independent from the OPC of the problem. To test whether the assumption on the selection rule is reasonable, we show this assumption holds for the well-known horse race rule in the second example. In the third example, we consider some other problem types and show $P_{success}$ of the horse race rule is still well above 0.5.

Example 1

For example, to cover a wide range of optimization problems, suppose there are N designs in total with true performances $J(\theta_i) = -(i/N)^a$, $i = 1,2,...N$, where $a>0$ is a parameter.[8] Fig. 7.15 below illustrates the fact that, by

---

[8] In (Lin 2000a) a maximization problem with true performances $J(\theta_i) = (i/N)^a$. Since we focus on minimization problem in this book, we reformulate the problem as a minimization problem. This is how the negative sign before $(i/N)^a$ comes.

using different values of "*a*", we can model a variety of ordered performance curves from neutral to steep. We are dealing with minimization problem, i.e., $\min_{i=1,2,\ldots N} J(\theta_i)$. Define the good enough set as designs within top-*g*% true performances, i.e., no greater than $(1-g\%) \times \min_{i=1,2,\ldots N} J(\theta_i) = g\%-1$. When *a* increases, the ordered performance curve changes from flat to steep. The size of G_v decreases. (See Fig. 7.15.) If we use the blind pick rule, $P_{\text{success}} = |G_v|/N$. The designs in G_v should satisfy $-(i/N)^a \le g\%-1$. So $i \ge \lceil N(1-g\%)^{1/a} \rceil$, where $\lceil \bullet \rceil$ is the ceil function. Thus $|G_v| = N-(\lceil N(1-g\%)^{1/a} \rceil-1)$ and $P_{\text{success}} = (N-\lceil N(1-g\%)^{1/a} \rceil+1)/N$. When *a* increases to infinity, P_{success} decreases to $1/N$. If *N* is usually extremely large, say no less than 10^8, P_{success} becomes meaningless. This means blind pick is not an efficient way to find the truly optimum, which is consistent with our experience and intuition.

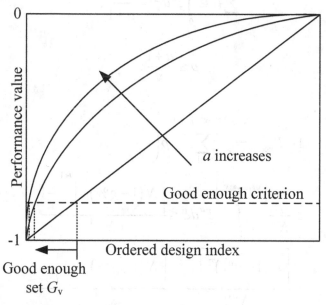

Fig. 7.15. Illustration of the reduction of good enough set when *a* increases

Fortunately, we usually have a crude model to estimate (though very roughly) the performance of the designs. Using a reasonable crude model, a truly better design should have a larger chance to be observed better, and thus have a larger chance to be selected by a selection rule that uses problem information. With this assumption, we can improve P_{success}. For example, to simplify the discussion, suppose a selection rule only selects one design instead of multiple designs each time. Denote the probability that a design is selected by the selection rule as $P_{\text{select}}(\theta_i)$. Assume $P_{\text{select}}(\theta_i)$ is proportional to the absolute value of $J(\theta_i)$, i.e., $P_{\text{select}}(\theta_i) = p(-J(\theta_i))$, where p is a positive constant.[9] Because only one design will be selected each time, we have

$$\sum_{i=1}^{N} P_{\text{select}}(\theta_i) = 1 \text{ and } P_{\text{select}}(\theta_i) = p\left(\frac{i}{N}\right)^a.$$

So, we have

$$p = \frac{N^a}{\displaystyle\sum_{i=1}^{N} i^a}.$$

Because

$$\sum_{i=1}^{N} i^a \geq \int_{0}^{N} \xi^a d\xi = \frac{N^{a+1}}{a+1},$$

we have

$$p \leq \frac{a+1}{N}.$$

Then

$$1 - P_{\text{success}} = \sum_{i=1}^{\left\lceil N(1-g\%)^{\frac{1}{a}}\right\rceil - 1} p\left(\frac{i}{N}\right)^a$$

$$\leq \frac{p}{N^a} \int_{1}^{\left\lceil N(1-g\%)^{\frac{1}{a}}\right\rceil} \xi^a d\xi \leq \frac{\left\lceil N(1-g\%)^{\frac{1}{a}}\right\rceil^{a+1} - 1}{N^{a+1}}$$

$$\leq \frac{\left\lceil N(1-g\%)^{\frac{1}{a}}\right\rceil^{a+1}}{N^{a+1}} \leq \frac{\left(N(1-g\%)^{\frac{1}{a}} + 1\right)^{a+1}}{N^{a+1}}$$

$$= \left((1-g\%)^{\frac{1}{a}} + \frac{1}{N}\right)^{a+1}.$$

---

[9] Admittedly, this is a rather specialized assumption. We make it here to illustrate the fact that additional structural information about a problem is needed if we need to make headway on the issue of performance value. Results of this section are rather preliminary and more of an anecdotal nature. More research is necessary.

When N goes to infinity, $((1-g\%)^{1/a}+1/N)^{a+1}$ converges to $(1-g\%)^{(a+1)/a}$ $<1-g\%$, so $P_{success}>g\%$. Thus in this example we obtain a lower bound of $P_{success}$ which is independent from the problem type (described by the value of a). We can use Eq. (7.27) to calculate the number of trials we need in order to find at least one truly good enough design (w.r.t. value) with high probability.

Example 2
Though this lower bound of $P_{success}$ is obtained under the assumption that the selection probability is proportional to the absolute performance value of designs, we find by simulation that it actually also holds for horse race rule, which are frequently used in real life problems. To see this, we do the following experiments. Assume the observation noise is i.i.d. $N(0,0.5)$, which is not a small noise since the range of performance value is only $[-1, 0]$. The performance function is still $J(\theta_i) = -(i/N)^a$ with $N = 1000$ and a ranges from 1 to 100. Define the good enough set as $G_v \equiv \{\theta_i \mid J(\theta_i) \leq -0.9\}$, i.e., $g\% = 0.1$. The success probability for the observed best design to be a good enough design is estimated from 10000 trials. From Fig. 7.16, we can see that the success probability of horse race rule is well above the lower bound for the range of a tested.

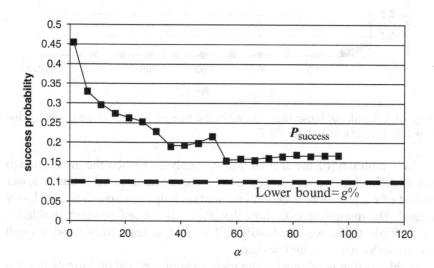

Fig. 7.16. Comparison between $P_{success}$ of horse race rule and the lower bound $g\%$ (Lin 2000a) © 2000 IEEE

Example 3

We also test this lower bound in other types of problems. Suppose the true performance of the designs is independently drawn from the standard normal distribution $N(0,1)$. This is a generic representation of many real world problems. For example, the distribution of tour lengths of traveling salesman problems is well modeled by a normal distribution (Stadler and Schnabl 1992). The observation noise is i.i.d. $N(0,0.5)$. The good enough set is defined as $G_v \equiv \left\{ \theta_i \middle| J(\theta_i) \leq -0.9 \min_{\theta_i} J(\theta_i) \right\}$, i.e., $g\% = 0.1$. The success probability and the average percentage of the good enough set in the overall design set are compared, with N ranging from 20 to 1000. Fig. 7.17 shows that the relative size of the good enough set, i.e., $|G_v|/N$, approaches 0 when N increases. On the other hand, the success probability floats well above 0.5.

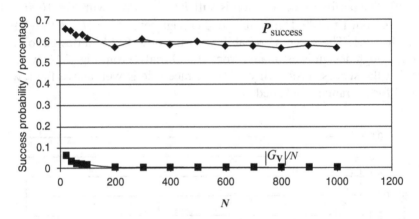

Fig. 7.17. Results of horse race rule when the true performances are normally distributed (Lin 2000a) © 2000 IEEE

As a summary of the second way to apply OO when the good enough set is defined in value, the key point is to estimate or to obtain a lower bound of $P_{success}$. More problem information helps to obtain a tighter lower bound. The numerical tests show that $P_{success}$ is not bad on many problems. In multiple trials, we can hardly fail to find at least truly good enough (w.r.t. performance value) designs.

In this section we discuss some ideas to apply ordinal optimization when the good enough set is defined w.r.t. to performance value instead of performance order. The presented ideas are worthy of consideration in real applications. In general, however, it is still an open question how to find a good enough design efficiently. Future studies on this problem are needed.

6 Combination with other optimization algorithms

As mentioned in Chapter II, ordinal optimization can be easily combined with other optimization algorithms. Since OO was developed in 1992, there have been many successful combinations with other optimization algorithms, such as genetic algorithm (GA), simulated annealing (SA), tabu search (TS), Lagrangian relaxation (LR), stochastic dynamic programming (SDP), hill climbing (HC), and nested partition, just to name a few. In this section, we discuss the basic ideas of these combinations, and use some numerical examples to show how the combination with other algorithms helps to improve the design quality and computationally efficiency.

The successful combination of OO with other optimization algorithms is mainly based on the following three ideas (Table 7.4). First, **regard other optimization algorithms as selection rules in OO**. OO can be regarded as a two-stage procedure to narrow down the design space. First from the entire design space to the set of N randomly sampled designs Θ_N, which hopefully is a reasonable representative of Θ, and then from Θ_N to the selected set S. Different from the heuristics and experienced based method to narrow down the search space, OO quantifies the confidence level (P, say no less than 0.95) of containing some good enough designs of the entire design space during this narrow down process. In many meta-heuristics-based optimization algorithms[10], such as genetic algorithm (Holland 1975; De Jong 1975; Goldberg 1989; Chambers 1995, 1999, 2000), simulated annealing (Kirkpatrick et al. 1983, Metropolis et al. 1953; Van Laarhoven and Aarts 1987; Aarts and Korst 1989), tabu search (Glover 1986, 1989, 1990; Glover and Laguna 1997), and ant colony optimizer (Dorigo 1992; Dorigo and Di Caro 1999; Dorigo and Stützle 2004; Dorigo et al. 1996, 1999), although we know the algorithm converges to global optimum asymptotically, we do not know the global goodness of the design thus found after finite number of iterations. By regarding these meta-heuristics-based optimization algorithms as special selection rules of OO, we can quantify the solution of this problem, and supply an easy way to determine the stopping criteria for these algorithms, which is a major concern in many practical applications of these algorithms. For example, in Subsection 6.1 we will report the work in (Zhang 2004) on the combination of GA and OO and the work in (Yen et al. 2004) on the combination of SA and OO.

Second, **apply OO to optimize the parameter of other algorithms.** The performance of some algorithms is sensitive to the parameter settings.

---

[10] Meaning general algorithms based on a set of heuristic principles.

For example, in genetic algorithms these parameters are the population size, the mutation probability, the cross-over probability, the mutation operation and cross-over operation; in simulated annealing these are the initial temperature, the temperature decaying ratio, and the inner number of iterations in each temperature for simulated annealing. Experts of these algorithms may know how to determine the appropriate parameter settings for different problems. In general, however, it is a difficult task. Since for most problems the only way to accurately evaluate the performance of an algorithm under a parameter setting is to use simulation, we can formulate a simulation-based optimization problem to find the appropriate parameter setting for an algorithm. OO is useful in this sense to find good parameter settings using short number of simulations. For example, in Subsection 6.2, we will report the work in (Zhang 2004) on the application of OO to find good parameter settings of GA and SA.

Third, **apply the basic ideas of OO**. OO justifies theoretically the two basic ideas, namely goal softening and ordinal comparison, which are pervasively adopted in engineering practices. There are various ways to apply these two ideas during the optimization process of other algorithms. For example, Mori and Tani combined OO and Tabu search (Mori and Tani 2003); Sullivan and Jacobson combined OO and hill climbing (Sullivan and Jacobson 2000); Luo et al. combined OO, GA, and linear programming to solve mixed integer programming problem (Luo et al. 2001); Luh et al. combined Lagrangian relaxation, stochastic dynamic programming, and OO in (Luh et al. 1999). Comparing with the first two ideas in Table 7.4, these combinations are more problem dependent. We will not discuss these combinations in details. Interested readers may refer to the corresponding papers for technical details, and (Shen and Bai 2005) for publications on other such combinations.

Table 7.4. Combination of OO with other optimization algorithms

Ideas	Examples
Other algorithms as selection rules in OO	GA+OO (Zhang 2004), SA+OO (Yen et al. 2004)
Apply OO to optimize the parameter settings of other algorithms	OO for GA and SA (Zhang 2004)
Apply the ideas of goal softening and ordinal comparison during the optimization process of other algorithms	Tabu search + OO (Mori and Tani 2003), Hill Climbing + OO (Sullivan and Jacobson 2000), GA + OO + linear programming (Luo et al. 2001), Lagrangian relaxation + OO + stochastic dynamic programming (Luh et al. 1999)

Before we proceed to the detailed discussion on the combinations, we need to make it clear that the above classification of the combinations is only a rough one and incomplete. It is still an open question how to combine OO with other optimization algorithms. We mention the above three ideas to give the readers some suggestions if they want to do such combinations. We do not want the readers to regard these ideas as restrictions.

6.1 Using other algorithms as selection rules in OO

Genetic algorithms (GA) and simulated annealing (SA) are known as useful meta-heuristics to solve optimization problems with large search space. One difficulty is to determine the stopping criteria. When we regard GA and SA as selection rules, we can use OO to quantify the total number of designs we need to sample in order to find some truly good enough designs with high probability[11]. In this way, we find an easy way to determine the stopping criteria. In subsection 6.1.1, we show the combination of GA and OO to deal with deterministic and stochastic optimization problems. Numerical examples on a flow job problem (defined later) shows promising results in terms of design quality. The concept of crude model supplies an easy way to incorporate problem information successively when the optimization process evolves. In subsection 6.1.2, we show the combination of SA and OO, and the use of a crude model based on neural networks, which can be easily updated during the optimization process. The performance of this method was demonstrated on a trim-loss problem.

6.1.1 GA+OO

In this subsection we mainly report the work in (Zhang 2004) on the combination of GA and OO. Genetic algorithm is invented by John Holland, and known as a meta-heuristic to simulate the natural selection process (Holland 1975). Although there are various implementations, the basic procedure of GA contains encoding, initialization, evaluation, and genetic operations (usually including copy, cross-over and mutation). In GA, each design is encoded by a string (e.g., binary, decimal or real), like the DNA in the nature. First a group of designs are randomly sampled from the design space and called the initial population. A fitness function is used to evaluate the goodness of these designs. The fitness function is usually (a modified version of the) objective function. The design with higher fitness

---

[11] The selection of population from old generation to new generation in GA is similar to the horse race selection rule, and cross over or mutation can be interpreted as broadening the sampling.

is selected with higher probability to enter the following genetic operations. Each time when a pair of designs is selected, the corresponding
strings are decomposed into small pieces. By copying and pasting these
pieces into new strings, we obtain a pair of new designs. To imitate the
mutation in the natural selection, some pieces of the strings are changed
with probability. This copy-crossover-mutation procedure continues until
we generate a required number of new designs. We compare the fitness of
all the new designs (sometimes also compare with a best-so-far design),
keep the top-n ones be the next population, and reject the rest. This new
population is evaluated, and the next population is produced through the
genetic operations. This procedure continues until the stopping criterion is
satisfied, say the number of designs thus explored exceeds a given number.

Ideas
In the practical application of GA, there are usually three difficulties: how
to choose an appropriate fitness function, how to determine the appropriate
stopping criterion, and how to deal with the time-consuming simulation-
based performance evaluation of fitness. The first is due to the fact that
the fitness function has a big impact on the explorability of GA. Since the
probability used in the genetic operations of traditional GA depends on the
evaluation of the fitness function, where a bad one could make GA stuck
to local minimum. The second is due to the fact that GA does not guarantee the global goodness of the solution thus found. It is a problem when to
stop the iteration. The third one of computational burden is common in all
the simulation-based optimization problems. The basic idea of GA+OO is
to apply the OO method to deal with these three difficulties.

For the first difficulty that the performance of GA is sensitive to the fitness function, the reason is that in traditional GA the value of the fitness
function is usually used to determine the crossover and mutation probability of each design in the population (Zhang 2004). OO suggests to regard
the original objective function as the true model, and the fitness function as
the crude model.[12] According to OO, we should select the observed good
ones instead of only the observed best one. Following this idea, in GA+OO
after we use the fitness function to evaluate the designs in the current
population, we keep the observed good ones, and reject the rest. These observed good designs are used in the cross-over and mutation to produce
the next population. Furthermore, the crossover and mutation probability

---

[12] The literature of GA does not provide guidelines on how to construct a fit
ness function from a performance function. However from the viewpoint of OO,
fitness can be considered as a crude model of the performance to ease the computational burden.

of these designs are set to be related to the observed order instead of the observed value of the fitness function, which makes the performance of GA+OO less sensitive to the fitness function.

For the second difficulty that GA does not guarantee the global good-ness of the design finally found, by regarding GA as a selection rule, GA+OO allows us to quantity the global goodness of the design(s) thus found. Recall that in Chapter II we have mentioned that if we blindly pick N designs from the entire design space, the probability that none of these N designs is within top-g% of the entire design space is $(1-g\%)^N$. Thus the probability that at least one of the N designs is truly top-g% is $1-(1-g\%)^N$. If we want this probability to be no less than P (say $P = 0.95$), we have

$$N \geq \left\lceil \frac{\ln(1-P)}{\ln(1-g\%)} \right\rceil .$$

(7.28)

It is usually believed (and reasonable to believe) that GA finds designs bet-ter than (or at least no worse than) blind pick. If we regard N as the designs sampled in the entire optimization process of GA, Eq. (7.28) calculates the upper bound of N, after the user defines the quality of the solution that s/he wants (i.e., how good must be the designs does s/he want (g%) and with how high a probability (P)). This can serve as both the stopping criterion and/or the size of the initial population depending on the approximate an-swer one wishes to get.

For the third difficulty of time-consuming simulation-based perform-ance evaluation, in GA+OO, a crude model is used in the place of per-formance function to separate the good from the bad with high confidence on the group of designs in each generation. Furthermore, many selection rules in OO help to allocate the computing budgets appropriately. For ex-ample, the population size in GA during each iteration is usually fixed. The OCBA (introduced in Section 4) can be used, instead of equally allo-cating the computing budgets to all the designs in the population as in the traditional GA.

These are the basic ideas of GA+OO. As a comparison, we list the basic procedures of both GA and GA+OO in Table 7.5, and use some numerical results on the flow shop problem in the next subsection to demonstrate the advantage of GA+OO, comparing with traditional GA.

Table 7.5. Comparison between the basic procedures of GA and GA+OO

GA	GA+OO
Encoding	Encoding
Initialization	Initialization. The size of the initial population is determined by Eq. (7.28).
Evaluation	Evaluation. When there are observation noises, a crude model is used, and the OCBA is used for allocating the computing budget.
Cross-over	Cross-over. The observed top-n designs in the population are selected for cross-over.
Mutation	Mutation. The observed top-n designs in the population are mutated with probability.
Update the population	Update the population
Check the stopping criteria.	Check the stopping criteria, which is the total number of designs explored during the process. This number is determined by Eq. (7.28).

Numerical results

Flow shop problem describes the scheduling problem in a manufacturing system. There are m machines and n parts. At any time, each machine can work on no more than one part, and each part can be machined by no more than one machine. The machining of a part takes some time. Once a part starts the machining on a machine, there should be no interruption until the machining finishes. There are specific procedures for each part. Each procedure describes the requirement on the order to use the machines. In a flow shop problem, the procedures of all the parts are the same. If some procedures are different, we have a job shop problem, which in general is more difficult to solve. To solve a flow shop problem, we need to tell each machine what order to follow, such that the time to finish all the parts (which is usually called the makespan) is minimized. In terms of our terminology, a schedule is a design and the makespan is our system performance. A flow shop problem with more than 3 machines is considered as an NP-hard problem, which means the problem is very difficult to solve.

(Zhang 2004) reported two groups of experiments. In the first group, the processing time of each part on each machine is deterministic, so the performance evaluation is easy, and we focus on comparing the robustness of GA and GA+OO w.r.t. problem instances. The comparison is over 29 benchmark problems (including the 8 testing problems car1, car2, ..., car8 from (Carlier 1978) and the 21 testing problems rec01, rec03,... rec41 from (Reeves 1995)), the data of which are available from OR-Library (OR-Library). The results are shown in Table 7.6, where C^* is the minimal makespan, the relative error (RE) is defined as $(C-C^*)/C^* \times 100\%$ with C

Table 7.6. Comparison on flow shop with deterministic processing time, where m is the number of machines, n is the number of parts, and RE is the relative error (Zhang 2004)

Problem instance			GA+OO			GA	
Problem index	n,m	C^*	BRE	ARE	WRE	BRE	ARE
Car1	11,5	7038	0	0	0	0	0.27
Car2	13,4	7166	0	0	0	0	4.07
Car3	12,5	7312	0	0	0	1.19	2.95
Car4	14,4	8003	0	0	0	0	2.36
Car5	10,6	7720	0	0	0	0	1.46
Car6	8,9	8505	0	0	0	0	1.86
Car7	7,7	6590	0	0	0	0	1.57
Car8	8,8	8366	0	0	0	0	2.59
Rec01	20,5	1247	0	0.04	0.16	2.81	6.96
Rec03	20,5	1109	0	0.0	0	1.89	4.45
Rec05	20,5	1242	0	0.21	0.32	3.82	2.63
Rec07	20,10	1566	0	0.79	1.15	1.15	5.31
Rec09	20,10	1537	0	0.35	1.17	3.12	4.73
Rec11	20,10	1431	0	0.91	3.07	3.91	7.39
Rec13	20,15	1930	0.26	1.08	1.66	3.68	5.97
Rec15	20,15	1930	0.10	1.23	2.21	2.21	4.29
Rec17	20,15	1902	0	2.08	3.21	3.15	6.08
Rec19	30,10	2093	0.14	1.76	3.01	4.01	6.07
Rec21	30,10	2017	1.44	1.64	3.12	3.42	6.07
Rec23	30,10	2011	0.85	1.90	3.08	3.83	7.46
Rec25	30,15	2513	1.31	2.67	3.74	4.42	7.20
Rec27	30,15	2373	0.97	2.09	3.58	4.93	6.85
Rec29	30,15	2287	1.88	3.28	5.95	6.21	8.48
Rec31	50,10	3045	0.43	1.49	2.59	6.17	8.02
Rec33	50,10	3114	0.61	1.87	4.05	3.08	5.12
Rec35	50,10	3277	0	0	0.33	1.46	3.30
Rec37	75,20	4951	2.46	3.41	4.30	6.56	8.72
Rec39	75,20	5087	1.63	2.28	3.24	6.39	7.57
Rec41	75,20	4960	2.30	3.43	4.69	7.42	8.92

the design found by the algorithm. BRE is the best RE, ARE is the average RE, and WRE is the worst RE.

From Table 7.6, we can see that for GA+OO, the designs thus found is very close to the truly best; the ARE does not exceed 3.5% on all the 29 test problems; the BRE and WRE are usually close, which means GA+OO is robust w.r.t. problem instances. Thus we conclude that GA+OO beats GA in the sense of design quality and robust w.r.t. to problem instances. In the above comparison, GA+OO and GA use the same initial population size, iteration number, and genetic operators (i.e., partially mapped crossover and SWAP in mutation). Their difference, as shown in Table 7.5, mainly lies in

(1) how the performance is evaluated and

(2) the probability of copying (for crossover and mutation).

For the first difference, GA+OO allows the user to use a crude model to roughly but quickly estimate the performance. In Table 7.6, since there is no observation noise, we do not use such crude model, so GA and GA+OO use the same performance evaluation technique in that example. However, when there is observation noise, GA tries to evaluate the fitness function accurately, but equally allocate the computing budgets among all the designs. GA+OO uses OCBA to allocate the computing budgets. In the following we will use another example to compare GA and GA+OO when there are observation noises (in Table 7.7).

For the second difference, GA sets the probability of copying (for crossover) proportional to the performance of the design (measured by the fitness function). GA+OO first rejects the observed bad designs, and sets the probability of copying exponential to the observed order of the design. e.g., the i-th design has a copying probability of $2^{l-i}/(2^l-1)$, where l is the total number of observed good designs.[13] In Table 7.6, since the processing time is deterministic, there is no observation noise. The only difference lies in the copying probability.

In the second group of experiments, suppose the processing time of each part on each machine is stochastic, following a uniform distribution. Now only a noisy performance observation is available, so we need to be careful to allocate the simulation budget among the designs in the population. Suppose we can run at most 2000 simulations in each iteration. As a comparison, we equally allocate the computing budgets to all the designs in the population in traditional GA, but use OCBA to do the allocation in GA+OO. Since the design found by each method is random in each run, (Zhang 2004) ran both GA+OO and GA 20 times in each of the 29 benchmark problems thus considered, and measured the best expected makespan (BEM), the average expected makespan (AEM), and the worst expected makespan (WEM). The numerical results are shown in Table 7.7.

---

[13] Unfortunately, as shown in (Zhang 2004), the selection of the values of l has a big impact on the performance of GA+OO. If the value of l is too small, GA+OO may be stuck at local minimum. If the value of l is too large, more bad designs are considered and evaluated in each iteration, and this decreases the efficiency of the algorithm. In Table 7.6, we set $l = 60$. However, it is still an open question how to determine the value of l a priori.

Table 7.7. Comparison on flow shop with stochastic processing time (Zhang 2004)

Problem instance			GA+OO			GA		
Problem index	n,m	C^*	BEM	AEM	WEM	BEM	AEM	WEM
Car1	11,5	7038	7038	7038.0	7038	7038	7038.0	7038
Car2	13,4	7166	7166	7187.0	7376	7166	7197.5	7376
Car3	12,5	7312	7312	7333.1	7399	7312	7345.8	7422
Car4	14,4	8003	8003	8003.0	8003	8003	8003.0	8003
Car5	10,6	7720	7720	7740.6	7779	7720	7768.8	7821
Car6	8,9	8505	8505	8521.3	8570	8505	8544.0	8570
Car7	7,7	6590	6590	6590.0	6590	6590	6590.0	6590
Car8	8,8	8366	8366	8366.0	8366	8366	8366.0	8366
Rec01	20,5	1247	1247	1250.6	1272	1249	1269.5	1326
Rec03	20,5	1109	1109	1113.5	1121	1111	1114.6	1128
Rec05	20,5	1242	1242	1244.6	1253	1245	1251.1	1275
Rec07	20,10	1566	1566	1583.3	1599	1568	1593.3	1650
Rec09	20,10	1537	1537	1564.6	1588	1543	1571.2	1605
Rec11	20,10	1431	1431	1461.1	1517	1442	1482.7	1551
Rec13	20,15	1930	1930	1961.8	2007	1966	1991.8	2022
Rec15	20,15	1950	1950	1980.1	2020	1973	2004.6	2069
Rec17	20,15	1902	1909	1948.2	1980	1954	1988.2	2042
Rec19	30,10	2093	2120	2144.5	2205	2135	2173.6	2218
Rec21	30,10	2017	2046	2065.7	2088	2061	2099.6	2142
Rec23	30,10	2011	2043	2056.7	2077	2049	2097.4	2150
Rec25	30,15	2513	2564	2597.7	2640	2595	2639.7	2671
Rec27	30,15	2373	2411	2445.3	2494	2423	2483.7	2531
Rec29	30,15	2287	2322	2383.9	2453	2380	2448.3	2507
Rec31	50,10	3045	3129	3167.4	3215	3131	3224.7	3245
Rec33	50,10	3114	3114	3167.4	3225	3140	3219.9	3233
Rec35	50,10	3277	3277	3298.3	3347	3284	3349.4	3370
Rec37	75,20	4951	5166	5241.3	5352	5276	5340.8	5368
Rec39	75,20	5087	5252	5313.8	5440	5298	5407.3	5486
Rec41	75,20	4960	5193	5258.1	5378	5227	5390.0	5414

In Table 7.7, the AEM shows that GA+OO has a better average performance than GA. Also, the AEM, BEM and WEM of GA+OO are close to each other.

In summary, both groups of numerical results show that the combination of OO with GA helps to improve the performance of GA.

6.1.2 SA+OO

In this subsection, we mainly report the work in (Yen et al. 2004) on the combination of SA and OO.

Ideas

Simulated Annealing is a stochastic method for solving combinatorial optimization problems based on ideas from statistical mechanics. The theory has been extensively developed (Collins et al. 1988) and has many applications to different problems as discussed in the literature (Johnston et al. 1989; Eglese 1990; Ku and Karimi 1991; Koulamas et al. 1994; Painton and Diwekar 1994; Falcioni and Deem 2000). The typical SA scheme contains: initialization of the reference design, trial design generation, performance evaluation, reference design updating, and temperature reduction (also shown in the left column in Table 7.8). In a typical SA scheme, a reference design is given initially. A "trial" design in the neighborhood of this reference design is generated. The objective function of this trial design is calculated. If it is lower than the objective function of the reference design, the reference design is replaced by the trial design. Otherwise, an exponential type transition probability is applied to determine whether the design should be updated:

$$
P_{update} = \begin{cases} 1, & J\left(\theta^t\right) < J\left(\theta^r\right) \\ \exp\left(\dfrac{-\left(J\left(\theta^t\right) - J\left(\theta^r\right)\right)}{T}\right), & \text{otherwise} \end{cases} \tag{7.29}
$$

Table 7.8. Comparison of the basic procedures of SA and SA+OO

SA	SA+OO
Initialize the reference design	Initialize the reference design
Generate a trial design	Generate a trial design
Evaluate the trial design	Crude-model based evaluation of the trial design
Update the reference design with probability, which is affected by the temperature.	Update the reference design
Reduce the temperature gradually	Reduce the temperature gradually
Generate a new trial design, and repeat the above process until the temperature decreases to a freeze value.	Generate a new trial design, and repeat the above process until the temperature decreases to a freeze value.

where T is the annealing temperature in SA, search procedure, θ^r is the reference design and θ^t is the trial design. The temperature is reduced gradually according to a predetermined annealing schedule to ensure that the system is not trapped in a local minimum. The traditional SA tries to obtain the accurate performance evaluation of the trial design, which is practically infeasible when the evaluation is simulation-based. OO suggests using a crude model instead, which saves the computing budgets. The basic produce of SA+OO is shown in Table 7.8. (Yen et al. 2004) applying this method to deal with a trim-loss problem. We briefly report their results in the rest of this subsection.

Numerical results

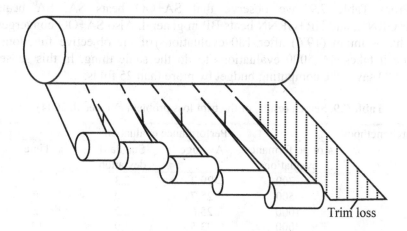

Fig. 7.18. A schematic illustration of the trim-loss problem (Yen et al. 2004)

The trim-loss problem appears when a set of ordered product reels are to be cut from raw paper reels or other reels with specified widths. The cutting process is simply a winding process, where the raw paper is wound through the slitter and cut by a set of knives positioned on the line, see Fig. 7.18. The product width can rarely be combined to the exact raw paper width, therefore there is a waste during the cutting. The main objective is to minimize the trim loss while demanded specifications are satisfied. To test the performance of SA+OO, Yen et al. considered a problem instance, the size of the design space of which is in the order of 10^{19}, and the best design of which is with performance 19.6 and was obtained by (Floudas et al. 1999).

Yen et al. have applied the SA+OO to deal with this problem. They use a generalized regression neural network (GRNN) (Specht 1991) as the crude model for performance evaluation. Furthermore, they apply this SA+OO procedure multiple times. Each time when the previous SA+OO procedure finishes, they record the reference designs that have been used, use longer time simulation to obtain more accurate performance evaluation of these designs, and update the neural network using this information.

This updated neural network will be used in the next iteration of the SA+OO procedure. Though it is reasonable to believe that SA is no worse than Blind Pick and Horse Race in general, it is of interest to find out how much better can SA be. They compare SA+OO only with SA, BP, and HR+GRNN, where HR+GRNN means to use Horse Race to find the observed good designs that are then used to update the GRNN. The numerical results are shown in Table 7.9.

From Table 7.9, we observe that SA+OO beats SA, SA beats HR+GRNN, and HR+GRNN beats BP in general. Also SA+OO converges to the optimum (19.6) after 180 evaluations of the objective functions, while it takes SA 5000 evaluations to do the same thing. In this sense, SA+OO saves the computing budget by more than 25 folds.

Table 7.9. Search results of the trim loss problem (Yen et al. 2004)

Search methods	Performance evaluation			
	# of performance evaluations	Average	Standard deviation	Time
BP	200	29.2	2.3	0.2
	500	25.7	3.0	0.7
	1000	25.6	2.2	2
	2000	23.5	1.9	12
	5000	22.4	0.8	82
HR+GRNN	200	23.3	1.3	13
	500	21.9	0.9	59
	1000	21.1	0.4	148
	2000	20.7	0.6	359
	5000	20.0	0.2	1873
SA	200	22.5	0.9	27
	500	21.5	0.9	81
	1000	20.1	0.7	113
	2000	19.8	0.2	262
	5000	19.6	0	493
SA+OO	120	22.4	1.5	7
	140	20.5	0.9	17
	160	19.7	0.2	31
	180	19.6	0	48
	200	19.6	0	58

6.2 Simulation-based parameter optimization for algorithms

As mentioned before, there is a common difficulty for many meta-heuristic algorithms, such as GA and SA, which is how to choose the appropriate parameter setting for a specific problem instance. For example, in GA some parameters can be tuned, such as the size of the population, the probability in the cross-over, and the probability in the mutation. There are also different ways to do the cross-over, e.g., partially mapped crossover (Goldberg and Lingle 1985), a two-point linear order crossover (Falkenauer and Bouffoix 1991), one-point crossover (Reeves 1995), and Non-Abel crossover. In SA, some parameters can be tuned, such as the initial temperature, the number of trial designs to sample under each temperature, and the annealing rate of the temperature. Giving an optimization problem, the performance of a meta-heuristic algorithm might be sensitive to the parameter settings, but in general we do not know which parameter setting will lead to the best performance. Of course, due to the heuristic nature of underlining optimization algorithms, there is generally no guarantee best solution that can be found even under the "optimal" parameter setting. However, it is still worthwhile to use suitable parameters because bad parameter settings will search inefficiently. Here OO can help. To get a reasonable setting for parameters, it would be too time consuming, if we test all the parameter settings of the same problem. Instead, one can formulate a two-level simulation-based optimization problem (Fig. 7.19). On the high level, each design is a possible parameter setting for the meta-heuristic algorithm, say GA. On the low level, we use the GA under the parameter setting specified on the high level to solve the given optimization problem. The solution on the low level problem will be used to measure the performance of the design (i.e., the parameter setting) specified on the high level. In other words, the low level optimization is a simulation to evaluate the performance of the design. The high level optimization problem is a simulation-based optimization problem, and we can use OO to solve this high level problem.

Since there is randomness during the optimization process of GA, we need to run the low level optimization many times to evaluate the performance of a design in the high level problem (i.e., a parameter setting of GA) accurately. OO suggests using a crude model instead, e.g., only a small number of iterations in the low level optimization. Also, the selection rule such as OCBA can be used to allocate the computing budgets appropriately in the high level problem. These are the basic ideas of applying OO to find the appropriate parameter setting of the meta-heuristic algorithms for a given problem. (Zhang 2004) demonstrated how to use this idea to determine the appropriate parameter setting of GA and SA in the flow shop problem. Interested readers may refer to (Zhang 2004) for the technical details and results.

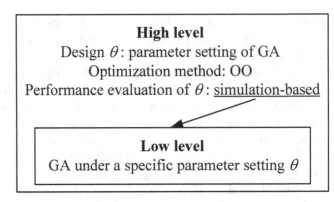

Fig. 7.19. Two-level simulation-based optimization for algorithms

6.3 Conclusion

In previous chapters, we claim that OO is a complementary tool for other optimization algorithms. In this section we further discuss possible ways to do the combination using some examples. These examples show that OO helps when used jointly with other heuristic algorithms in a reasonable way. Of course, we do not mean they are the only and the best way to do the combination. Up until now, to give the readers some guidelines in the application, we have mentioned three ideas to do the combination: regarding other algorithms as selection rule in OO, applying OO to find the appropriate parameter setting of the algorithm for a given problem, and applying the ideas of ordinal comparison and goal softening during the optimization process of other algorithms. It is still an on going topic how to combine OO with other algorithms. We hope readers to regard the above ideas only as suggestions, but not restrictions.

Finally, it is important to keep in mind that OO is complementary to and not meant as substitute for traditional optimization methods. Furthermore, despite its many successes, OO is not at all useful for the needle-in-the-haystack type of problems where nothing but the best will do. Thus, it cannot solve NP-hard combinatorial problems (although is can get close to the best in "order"). Besides, proximity in order does not imply proximity in value. Additional structural information on the problem are needed if we want good enough solution in terms of performance value as discussed in Section 5 above.

Chapter VIII Real World Application Examples

In the previous chapters we have introduced the methods of OO for single-objective unconstrained optimization (Chapter II), multi-objective optimization (Chapter IV), constrained optimization (Chapter V), and for simple and good enough strategies (Chapter VI). The purpose of this chapter is to demonstrate how these methods can be applied to real world problems. We consider four problems: three real world applications and a benchmark problem in team decision theory. In Section 1, we consider a scheduling problem for apparel manufacturing problem (Lee 1997, Bouhia 2004). We demonstrate how the OO method introduced in Chapter II helps to save the computing time by 2000 folds comparing with the brute force method in this problem. In Section 2, we consider a turbine blade manufacturing process optimization problem (Yang 1998 and Yang et al. 1997). The objective function in this problem is not stochastic simulation but deterministic complex calculation. In Section II.6 we have mentioned that OO can also be applied in this type of problem. We justify this through this application. Furthermore, we show how to obtain better estimate of the size of the selected set through appropriate interpolation when the noise level is different from the three values considered in the UAP table in Section II.5. In Section 3, we consider a remanufacturing system performance optimization problem (Song et al. 2005a, 2005b). There are constraints in this problem. We first demonstrate how to use COO introduced in Chapter V to deal with this problem directly. Then to better describe the requirements in engineering practice, we reformulate the problem to a two-objective optimization problem, and use VOO introduced in Chapter IV to solve the problem. We also demonstrate how we can incorporate the problem information to obtain less conservative estimate of the size of the selected set when the noise level is different from the values considered in the VOO-UAP table in Section IV.2. In Section 4, we consider the Witsenhausen problem, which is a famous problem in team decision theory and has not been solved for nearly forty years since the problem was first proposed by Witsenhausen in 1968 (Witsenhausen 1968). This is a strategy optimization problem, which has an extremely large search space. We demonstrate how OO helps to discover properties of the good strategies, thus successively narrows down the search space, and substantially improves the strategy that has been obtained before the application of

OO to this problem. Based on the properties thus discovered, Lee et al. obtained the best-so-far strategy for this problem (Lee et al. 2001). We also demonstrate how to use the OBDD introduced in Chapter VI to quantify the complexity of the milestone solutions to this famous problem. Combining OO with OBDD, we demonstrate how to search for a good and simple strategy in this problem.

As in all cases, we present only the salient features of the problem, the methodology used, together with enough results to show the improvement obtained and/or savings achieved. Readers can consult the original references for minute details.

1 Scheduling problem for apparel manufacturing

In this section, we apply the OO method in Chapter II to solve a scheduling problem for apparel manufacturing which are subject to the whims of fashion. The manufacturing system is characterized by the co-existence of the two production lines, i.e., one with long lead time and low cost, the other a flexible one with short lead time and high cost. The goal is to decide: (1) the fraction of the total production capacity to be allocated to each individual line, and (2) the production schedules so as to maximize the overall profit and yet avoid stock shortage. The problem is difficult and it is prohibitive to search for the best solution in view of the tremendous computing budgets involved. Using ordinal optimization introduced in Chapter II, we have obtained very encouraging results – not only have we achieved a high proportion of "good enough" designs but also tight profit margins compared with a pre-calculated upper bound. There is also a saving of at least 2000 folds of the computation time if brute-force simulations were otherwise conducted (Lee 1997). The rest of this section is organized as follows. In Section 1.1, we introduce the background of this problem. A detailed problem formulation is presented in Section 1.2, together with a discussion on the challenges to solve the problem. Section 1.3 introduces the application of ordinal optimization in this problem, including how to randomly sample designs and how to construct a crude model, which is computationally fast, evaluating the performance of the designs roughly, and useful to compare the designs. The application procedure of OO in this scheduling problem is also summarized. Section 1.4 gives several experimental results to show how ordinal optimization helps to save the computing budgets by 2000 folds, and the performance of the good enough design found by OO is close to the upper bound of the performance of the optimal design. We make a brief conclusion in Section 1.5.

1.1 Motivation

In the past thirty years, technological advancements, international competitions and new market dynamics have had major impacts on the North American apparel manufacturing industry. The conventional analysis of the apparel industry predicts that the apparel industry will collapse rapidly and migrate to nations with low labor costs. Although apparel industries still exist in the United States nowadays, intense competition encourages management to develop new production and supply methodology in order to remain competitive (see (Harvard 1995)). One key issue involved is the allocation of scarce production resources over competing demands. Before we introduce the detailed scheduling problem from the manufacturer's viewpoint in the next subsection, let us first have a big picture of the entire apparel manufacturing system. A typical apparel manufacturing system is shown in Fig. 8.1. The retailers receive customer demands, which is usually random and sometimes have seasonal variations. For example, the demands on swimsuits are high in spring but low in fall, while the demands on ties are high on Father's day and Christmas but low in the rest of the year. It is an important strategic requirement to satisfy customer demands. Failing to do so can result not only in lost profits due to reduced sales, but also the lost of future market share. In order to deal with the randomness in the customers' demand, the retailers usually maintain a small inventory of the apparel. In the apparel market nowadays, customers demand the variety of products. Thus the retailers have responded to their customers' wishes by maintaining a small inventory of many different styles of apparel and demanding rapid (usually weekly) replenishment of store inventory from the apparel manufacturers.

From the manufacturer's viewpoint, to fulfill the random replenishment orders from the retailers, an inventory of finished goods is established.

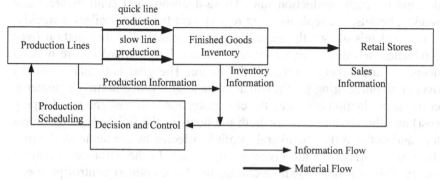

Fig. 8.1. Material and information flow chart of apparel manufacturing systems

This inventory is expensive to be maintained and should be no higher than necessary to meet demands. There are generally two ways to build up a required inventory level, i.e., using quick or regular production lines. In a regular production line, work flows from worker to worker in bundles with work buffers between each work station. The Work-In-Process (WIP) in each buffer is so large that it takes 20 to 25 days for a garment to pass through all operations, even though only 10 to 20 minutes of direct labor content is actually required to assemble the garment. Therefore, this kind of line has a long lead time, which is defined as the time from the receipt of the order from the retailers to the time the products are delivered to the finished goods inventory, i.e., the order is fulfilled. In a quick line, a small group of workers are cross trained to perform several sewing operations. The group of workers performs all of the sewing assembly operations on the apparel item. Workers move from one work station to another thereby minimizing the WIP in the production line. The cycle time in the quick line is less than the regular line; however, since a cross-trained worker is more expensive than a regular worker, and workers are generally less productive, on average, at several operations than they are at a single operation, the cost of the quick line is higher. The manufacturer is responsible for making decisions on how to manage future production on different production lines in order to maximize the overall manufacturing profit. We discuss more details of the manufacturer's scheduling problem in the next subsection.

1.2 Problem formulation

The manufacturer's decision and control should determine (1) the fraction of the total production capacity, γ, to be allocated to each production line; and (2) the scheduling strategy, α, that decides when to work on which demand on each production line. These decisions and controls are made weekly, because the replenishment requirement from the retailers is weekly-made, and this allows the manufacturer to collect new information (past production schedules, inventory and demand information) before making decisions and having control of the future. The goal is to maximize the overall manufacturing profit, which is the total revenue minus the material cost, the production cost (i.e., the cut, make, and trim cost and the shipping cost) and the holding cost for both the items in the finished goods inventory and in the WIP. The overall profit is affected by the demand, the production facilities, and the inventory dynamics. In the parlance of control theory, this is a full blown stochastic feedback optimal control problem. We will discuss these issues separately in the rest of this subsection. Before that, we need to introduce the concept of Stock Keeping Unit (SKU), which is used to describe different types of items in the demand and the production. A SKU is a particular style, fabric and size of an apparel item. A typical jeans manufacturer may make 10,000 to 30,000 distinct SKUs of jeans in a year. In a given season of the year, the number of SKUs manufactured may still be as high as 10,000. In our problem, suppose there are M SKUs in all.

1.2.1 Demand models

We assume that demand is weekly-made and there is no back-ordering, i.e., the retailers will be given whatever is left in the warehouse when the demand level is greater than the inventory level for each SKU. Assume the demand of SKU i at time t, $d_i(t)$, is max($\zeta(t)$,0), where $\zeta(t)$ is a Gaussian random variable with mean $\mu_i(t)$ and standard deviation $\sigma_i(t)$. If we neglect the truncated effect, average demand of SKU i at time t is equal to $\mu_i(t)$. When we consider seasonal effects, $\mu_i(t)$ will be a periodic function. Coefficient of variation of SKU i, $Cv_i(t)$, is defined as standard deviation divided by mean, i.e.,

$$Cv_i(t) = \frac{\sigma_i(t)}{\mu_i(t)}. \qquad (8.1)$$

We assume that the coefficient of variation, $Cv_i(t)$, is a constant, and we will use Cv_i from now on.

There are usually three types of demands in apparel manufacturing system.

- Flat Demand

There is no consideration on seasonal effects. The average demand is constant throughout the year, i.e., $\mu_i(t)$ = constant. Fig. 8.2 shows the relationship between the average demand and real demand under different Cv's.

- Seasonal – Sine Demand

There is a consideration on seasonal effects and the changes of the average demand are smooth, i.e., $\mu_i(t) = A_i + B_i\sin(2\pi t/T)$ where T is the period of seasonal effects, and A_i, B_i are the amplitudes of the function, $A_i > B_i$. Fig. 8.3 shows the relation between the mean and the real demand under different Cv's.

- Seasonal – Impulse demand

Sometimes, there are peak demands caused by promotions or special holidays or both. A sudden jump in sales is often observed at the beginning of a peak sales period. The peak sales are often planned to be roughly equally spaced along a year and last for a short time (several weeks) compared with the regular selling period. The seasonal demand can be modeled by a two-level demand function, which is called impulse demand. Fig. 8.4 shows the relationship between the mean and the real demands under different Cv's.

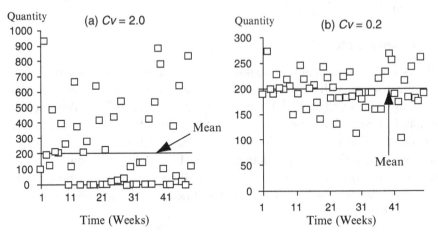

Fig. 8.2. Actual demand vs. mean for flat demand case when (a) $Cv = 2.0$ and (b) $Cv = 0.2$

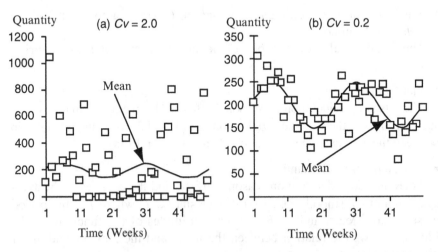

Fig. 8.3. Actual demand vs. mean for seasonal – sine demand case when (a) $Cv = 2.0$ and (b) $Cv = 0.2$

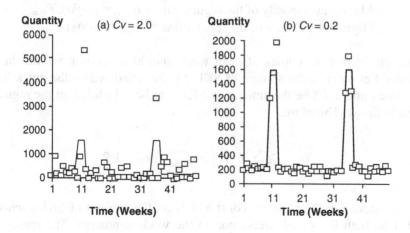

Fig. 8.4. Actual demand vs. mean for seasonal – impulse demand case when (a) $Cv = 2.0$ and (b) $Cv = 0.2$

1.2.2 Production facilities

As aforementioned, we consider two different kinds of production lines, a quick line and a regular (slow) line; the lead times of which are denoted respectively as L_q and L_s. Both L_q and L_s are assumed to be known and constant. By definition, $L_q < L_s$. The total production capacity is generally limited by the availability of resources such as equipments. In the apparel industry, production capacity is generally determined by available labor. In this problem, we assume the total capacity CP is equal to the weekly average demand over all SKUs. Regularly, there are 5 working days a week, and we allow one day overtime, therefore,

$$\text{Maximum capacity} = CP_{max} = 1.2 \times CP.$$

As for the minimum capacity, it is clear that it should be at least greater than zero, but in most situations, it cannot vary greatly week to week. A reasonable assumption is we have to work at least 4 days a week. Therefore,

$$\text{Minimum capacity} = CP_{min} = 0.8 \times CP.$$

The ratio of the quick line capacity to the total capacity, γ, is a constant to be determined. Therefore,

$$\text{Maximum capacity of the quick line} = u_{1max} = \gamma \times CP_{max},$$
$$\text{Minimum capacity of the quick line} = u_{1min} = \gamma \times CP_{min},$$

Maximum capacity of the regular line = $u_{2max} = (1-\gamma) \times CP_{max}$,
Minimum capacity of the regular line = $u_{2min} = (1-\gamma) \times CP_{min}$.

The production schedules of each week should be chosen within these limits. Let $u_{i1}(t)$ be the amount of SKU i to be scheduled on the quick line at time t and $u_{i2}(t)$ be the amount of SKU i to be scheduled on the regular line at time t. Therefore,

$$u_{j\max} \geq \sum_{i=1}^{M} u_{ij}(t) \geq u_{j\min} \text{ for } j = 1,2.$$

Please note that the capacity constraint is on the amount of SKUs scheduled on both lines each week, not on the work-in-process. The reason is that, in an apparel manufacturing system, the time to directly produce the items is comparatively smaller than the waiting time, which is also the main cause of the lead time. We assume if the capacity constraint is satisfied when scheduling the SKUs, then it is always possible to manage the workers and the machines to finish the scheduled SKUs within the lead time of that production line, although the more SKUs are scheduled, the higher the WIP will be, and this will increase the holding cost for the WIP.

1.2.3 Inventory dynamic

There is a weekly replenishment from the finished goods inventory to the retail stores. Let $I_i(t)$ be the total inventory of SKU i at time t, and $W_i(t)$ be the total work-in-process (WIP) inventory of SKU i at time t. Then $I_i(1)$ is the initial inventory of SKU i, and $I_i(t+1)$ should be equal to the left inventory level in the last week after satisfying the demand, i.e., $\max(I_i(t)-d_i(t),0)$, plus the SKUs that are produced during the last week by both production lines, i.e., $\sum_{j=1}^{2} u_{ij}(t-L_j+1)$. So we have

$$I_i(t+1) = \max\left(I_i(t)-d_i(t),0\right) + \sum_{j=1}^{2} u_{ij}(t-L_j+1), \forall i = 1,...,M.$$

The WIP of week t will be the sum of all the SKUs that are still in production, i.e.,

$$W_i(t) = \sum_{j=1}^{2} \sum_{k=t-L_j+1}^{t} u_{ij}(k).$$

1.2.4 Summary

For each product, let C_m be the material cost, C_{L_j} the production cost if line j is used, and P_S the sales price. Let also C_I be the holding cost for the finished goods inventory and WIP per week and per product. For a given γ and α, $J_{\text{total}}(\alpha,\gamma)$ denotes the total manufacturing profit gained from week $t=1$ to week $t=\Pi$, which is calculated as follows.

$$
\begin{aligned}
J_{\text{total}}(\alpha,\gamma) = & \sum_{i=1}^{M}\sum_{t=1}^{\Pi} P_S \min\left(I_i(t), d_i(t)\right) && \text{sales price} \times \text{sales} \\
& -\sum_{i=1}^{M}\sum_{t=1}^{\Pi}\sum_{j=1}^{2} C_m u_{ij}(t) && \text{material cost} \times \text{production} \\
& -\sum_{i=1}^{M}\sum_{t=1}^{\Pi}\sum_{j=1}^{2} C_{L_j} u_{ij}(t) && \text{production cost} \times \text{production} \\
& -\sum_{i=1}^{M}\sum_{t=1}^{\Pi} C_I I_i(t) && \text{inventory cost} \times \text{inventory} \\
& -\sum_{i=1}^{M}\sum_{t=1}^{\Pi} C_I W_i(t) && \text{inventory cost} \times \text{WIP.}
\end{aligned}
$$

The average weekly manufacturing profit, $J(\alpha,\gamma)$ is given by $J_{\text{total}}(\alpha,\gamma)$ divided by Π, i.e.,

$$
J(\alpha,\gamma) = \frac{1}{\Pi} J_{\text{total}}(\alpha,\gamma) \tag{8.2}
$$

Since the demand is random, the objective function of the scheduling problem is then to find γ and α in order to maximize the expected total manufacturing profit, i.e.,

$$
\max_{\alpha \in \Phi, \gamma \in [0,1]} E\{J(\alpha,\gamma)\},
$$

where Φ is the collection of all possible scheduling policies α.

There exist four nearly insurmountable challenges in this problem.

- First, in the apparel manufacturing system, different sizes, colors, or fashions of shirts are considered as different ***stock-keeping units***

(SKUs). There may be over ten thousands different SKUs in the system. The demand of each SKU varies weekly and exhibits seasonal trends.

- Second, since the exact demand is unknown in advance, in order to estimate precisely the expected profit of each strategy, one needs to perform numerous time-consuming and expensive Monte-Carlo simulations.
- Third, the number of applicable strategies is equal to the size of the possible production schedules raised to the power of the size of the information space. It is clear that this can be very large even for a moderately-sized problem.
- Fourth, since the neighborhood structure in the strategy space is not known and the performance value function cannot be explicitly represented in terms of strategy, the calculus and gradient decent algorithm cannot be applied.

Because of these difficulties, if we want to get the optimal solution to this problem, brute-force simulation, or large state-space dynamic programming is unavoidable. In practice where there are many different SKUs, it is computationally infeasible to find the global optimum. In the next subsection, we will apply OO to solve this problem and provide results which are not only good but also quantifiable.

1.3 Application of ordinal optimization

Since the number of SKUs is very large, the number of applicable strategies is an astronomically large number and Monte Carlo simulation is needed to evaluate the performance value of each strategy. Therefore, searching for the best solution is prohibitive in view of the tremendous computing budgets involved. As mentioned in Chapter II, if we do not insist on getting the optimal design, i.e., we soften our goal by having a high probability of getting any good enough design, the problem will become more approachable. When the goal is softened, we can tolerate imprecise performance estimates because we can have high confidence in obtaining a "good enough" design from a selected set. In this way the difficulties of the original problem can be overcome.

To apply ordinal optimization in this scheduling problem, two important questions must be answered: How can we randomly sample designs from the design space? What is the crude model that is computationally fast and can supply rough performance estimate? We discuss these two questions in turn.

1.3.1 Random sampling of designs

Each design is defined as (α, γ), where α is production schedule and γ is the ratio of the quick line capacity to the total capacity. Although samples of γ can be easily generated by a uniform random number generator, the random sampling of the production schedule α is not so straightforward. By definition, the production schedule should satisfy all the capacity constraints each week and make the level of the finished goods inventory track the demand in an appropriate way, so that most of the demands can be fulfilled and the inventory holding cost is reasonable. If we do not utilize the above information, but uniformly randomly sample production schedules that satisfy the capacity constraints, there is no reason to believe these schedules can track the demand and reduce the inventory holding cost appropriately. We should incorporate the above information in the random sampling of production schedules.

Suppose there is no uncertainty in the demand process $d(t)$, i.e., $d(t)=E[d(t)]$, and we can take $E[d(t)]$ to be a deterministic process, then we can arrange the production schedules to track $E[d(t)]$ as best as we can (see Fig. 8.5 (a)). This can be solved, in principle, by using well-known control theory tools such as dynamic programming, or other ad hoc heuristic methods, if the size is too large. However, as shown in Fig. 8.5(b), if the

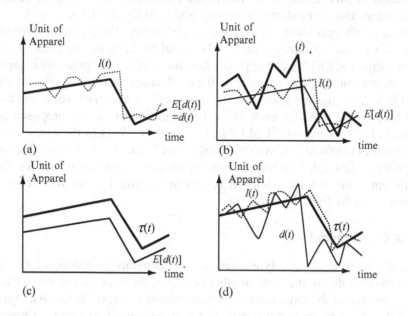

Fig. 8.5. A graphical illustration of how to select the scheduling strategy

demand process $d(t)$ is a random process, it is clear that tracking $E[d(t)]$ alone will not be satisfactory (in Fig. 8.5(b), we can see that the inventory $I(t)$ is too low to guarantee sales). Thus, we introduce another process to play the role of a deterministic process from which we can plan our scheduling strategy. This new process is called the target level, denoted as $\tau(t)$, which is used to replace what we have to, but cannot, track, i.e., $d(t)$. This is shown in Fig. 8.5(c). Notice that $\tau(t)$ is not a random process. Now we can solve a control problem to determine $u(t)$ to follow $\tau(t)$ as best as we can. $u(t)$ will be the production schedules. Therefore, we first find the target level of each SKU, then generate the production schedules that can track the target level. This is shown in Fig. 8.5(d). The remaining problem is to find a method to generate appropriate target levels and also a production schedule that will track the target level.

We must have higher inventory level when we have higher average demand, and also, when we have higher uncertainties, we must increase the inventory level in order to guarantee no shortage. So, a reasonable and simple way to generate the target level for each SKU is to let the target level of SKU i be proportional to the mean value and the standard deviation of the demand in the future, i.e., $\tau_i(t) = (a_1 + a_2 C_{vi})\mu_i(t + a_3)$, where $\mu_i(t)$ is the mean of the demand of SKU i at time t while Cvi is the coefficient of variation of SKU i, and a_1, a_2, and a_3 are constants constituting the design parameters and are randomly generated and used for all SKUs.

When the target level is given, we should arrange the production schedules so that the inventory level will be equal to the target inventory level by the time the SKUs exit the production lines. When the production capacity is not enough, the capacity will be allocated "fairly" among all the SKUs so that after the allocation the ratio of the inventory level to the target level is the same for each SKU. This algorithm was first proposed by (Bo et al. 1994) and modified by L. H. Lee (Lee 1997) to the cases when there are multiple SKUs, multiple production lines, and limited production capacity. Although this target tracking strategy does not guarantee the optimum, from our experimental results in Section 1.4, we will observe that it is not far from the optimum.

1.3.2 Crude model

The difficulty to accurately evaluate the performance of a design (α, γ) by simulation is due to the large number of SKUs, the long simulation in each replication, and the large number of replications in total. To obtain a crude model, which is computationally fast and only need to supply a rough performance estimate, we can do the following three relaxations. (1) As aforementioned, there might be 10,000 or even 30,000 different SKUs.

Instead of tracking the dynamics for each SKU during simulation, which will cost a lot of memory space and time, we aggregate the SKUs by the coefficient of variation. The mean of the demand of the aggregated SKU is equal to the sum of the means of the SKUs with similar Cv, and the Cv of the aggregated SKU will be equal to that Cv. In fact, if the demands of the SKUs are all positively correlated, it would be clear that this aggregation is appropriate. The reason is that, if $X_1, X_2, \ldots X_N$ are random variables with the same Cv and the correlation between X_i and X_j is equal to 1, $Y = \sum_{i=1}^{N} X_i$ will be a random variable with the same Cv and the mean demand is $\sum_{i=1}^{N} E[X_i]$. For other cases (say the demands of all the SKUs are independent, experiments have been done to justify this aggregation method (Lee 1997)). After the aggregation, there are usually no more than 100 SKUs and sometimes no more than 10 SKUs, which saves a lot of simulations. (2) Instead of simulating the system for several hundreds or thousands weeks, we can use a short simulation of only 100 weeks. (3) We can use a small number of replications (even only one replication).

In this way, we obtain a crude model. Although the performance estimate might be very different from the true performance values, the observed good enough designs set will nevertheless contain a lot of truly good enough designs.

Let us summarize the application procedure of OO in this scheduling problem (Box 8.1).

Box 8.1. Application procedure of OO in the scheduling problem of apparel manufacturing system

Step 1: Randomly generate N target levels as described in section 1.3.1.
Step 2: For each target level, randomly generate the capacity allocations between the two production lines. Then use the target tracking strategy to determine the production schedules of all the SKUs. The capacity allocation together with the production schedule is a design (α, γ).
Step 3: Aggregate the SKUs by the coefficient of variation. Then use the crude model to roughly estimate the performance of the designs.
Step 4: Estimate the observation noise level and the problem type.
Step 5: The user defines the size of good enough set g, and the required alignment level k.
Step 6: Use the UAP table in Section II.5 to calculate the size of the selected set S.
Step 7: The OO theory ensures that there are at least k truly good enough designs in the observed top-s designs with high probability.

1.4 Experimental results

In this subsection, we will present the experimental results in two experiments. First, we consider the case with 100 SKUs and show that the methods based on OO, as introduced in Section 1.3, can save the computing budgets by 2000 folds. Second, we modify the objective function to consider the requirement on satisfaction rates. The satisfaction rate is defined as the fraction of the time that the demand is satisfied by the finished goods inventory level. The experimental results demonstrate how the OO based method can be used as a platform to study the impacts of different factors on the total profit.

1.4.1 Experiment 1: 100 SKUs

There are 100 different SKUs. For each SKU, the demand at time t is a truncated Gaussian random variable with mean equals to $\mu(t)$ and coefficient of variation equals to Cv, i.e., $d(t)=\max(\zeta(\mu(t), Cv\mu(t)),0)$. We use seasonal-sine demand introduced in Section 1.2 to model the average demand $\mu(t)$. The ratio of the average demand from the peak season to the low season ranges from 3 to 7. The Cv of the SKUs ranges from 0.1 to 1.0, and the SKUs with higher Cv have lower demand than the SKUs with lower Cv. The ratio of the demand of the SKU with the highest Cv to the demand of the SKU with the lowest Cv is 5. The period of a season is 25 weeks, i.e., about half a year. We have 2 production lines, the lead time of the quick line is 1 week, while that of the regular line is 4 weeks. The weekly total production schedules should be maintained within 100±20% of the total production capacity. The "good enough" set G is defined as the top 5% of the solution space. We use the linear method introduced in Section 1.3 to generate the target inventory level, i.e.,

$$\text{target } \tau_i(t) = \left(a_1 + a_2 Cv_i \right) \mu_i \left(t + a_3 \right).$$

In order to get the *true* performance value of a design, it will be necessary to run the detailed simulation. In this experiment, we assume that a detailed simulation utilizes the entire 100 SKUs with a simulation time = 500 weeks and the number of replications = 40.[1] When we estimate the *observed* performance value of the design, we run an aggregated 10 SKUs simulation

---

[1] In fact, when we run the simulations, it takes about a week to run on a Sun SPARC 20 station. The estimated performance values are still imprecise, but the errors are very small (the standard deviation of the error is about 0.05% of the performance value).

with time = 100 weeks and number of replication = 1.[2] Notice that the time needed to estimate the *observed* performance value is roughly *1/2000* of the time to estimate the *true* performance value of the design. We have reduced the computation time from 1 week to several minutes.

Table 8.1. The cost structure of the shirt manufacturer

Cost term	Value
Inventory holding cost per unit per week C_I (both finished good and WIP)	$0.08
Quick line production cost per unit C_q	$4.4
Regular line production cost per unit C_s	$4
Material cost per unit C_m	$10
Sale price per unit P_s	$20

The cost structure is shown in Table 8.1. The results of the simulations are shown in Table 8.2.

Table 8.2. The alignment level and profit that we obtained when OO was used for the 100 SKUs case (periodic-sine demand)

s	k	J
1	1	356,834
5	4	358,999
10	7	358,999
20	11	358,999
50	26	359,504
100	38	359,504

In Table 8.2,

- s = number of designs selected by using the observed performance value.
- k = the average number of overlaps of the selected s designs with true top-50 designs, i.e., alignment level $|G \cap S|$. (These top-50 designs are obtained by running all 1000 designs for detailed simulation.[3])
- J = the best performance value (profit) in the selected s designs.

---

[2] From the simulation results, errors are about 3 to 4% of the performance value.

[3] Notice that this is a tremendous computational burden and precisely what our approach is trying to circumvent. However to lend credibility to our approach, this is the only way to prove its validity. Once established, we need not repeat this validation process in practical applications.

To get an idea of the absolute difference between the results obtained by OO and the true optimum, an upper bound of the profit can be obtained (Lee 1997). The idea is to consider a long enough simulation so that the system achieves the steady state. The average weekly production will be roughly equal to the average weekly sale; the ratio of the average weekly production of the quick line and the regular line should be close to the ratio of the capacity allocated to the quick line and the regular line (we can always fully utilize the capacity); and by Little's Law, the WIP should be equal to the average weekly production multiplied by the lead time. We will not deduce the upper bound in details. Please refer to (Lee 1997) for specific details. Based on these observations, the upper bound of the optimum profit is $369,551.

From the results in Table 8.2, we can make the following observations.

- In order to get the *true* performance value[4] of all the designs, the simulations were run about one week, 24 hours a day, on a Sun SPARC 20 machine, but to get the *observed* performance values, we only needed a run of several minutes.
- The selected set S contains a high proportion of good enough designs. When we increase the size of selected set S, the number of alignments between the good enough set and the selected set S also increases.
- The performance value (manufacturing profit) of the best design in the selected set is indeed very close to the pre-calculated upper bound (3% from the upper bound), which means that this approach not only guarantees to find good designs but also the design is close to the optimum in this problem.

1.4.2 Experiment 2: 100 SKUs with consideration on satisfaction rate

For some companies, it is an important strategic requirement to satisfy customer demands. Failing to do so can result in not only the lost profits due to lost sales, but also the loss of future market share. This motivates a concept called the *satisfaction rate*, which is simply the fraction of the time that the demand is satisfied by the inventory level. A satisfaction rate of 1 means that customer demands will always be fulfilled from inventory, or in other words, the inventory level is higher than the demand level every week. Satisfaction rate is defined as,

---

[4] The true performance values are obtained by running detailed simulation.

$$\text{satisfaction rate} = \frac{1}{\Pi} \sum_{t=1}^{\Pi} \iota\big(I(t) - d(t)\big)$$

Where

$$\iota(x) = \begin{cases} 1 & \text{if } x \geq 0 \\ 0 & \text{if } x < 0 \end{cases}.$$

Therefore, in order to maintain a high level of satisfaction rate, it is unavoidable to keep a high inventory level, which will induce a cost. However, the relation between enforcing the satisfaction rate and the cost incurred is not obvious. In this section, by using the OO based method in Section 1.3, we can quickly find this relation, which will serve as a good indicator for the production and sales managers to know how to set their satisfaction rate level.

Assume that the satisfaction rate constraint is that the average satisfaction rate, *SR*, of all SKUs have to be above certain level, β, i.e.,

$$SR = \frac{1}{M\Pi} \sum_{i=1}^{M} \sum_{t=1}^{\Pi} E\big[\iota\big(I_i(t) - d_i(t)\big)\big] \geq \beta.$$

Therefore the scheduling problem becomes

$$\max_{\alpha \in \Phi, \gamma} E\{J(\alpha, \gamma)\} - \text{Penalty}\,(SR; \beta)$$

subject to the constraints on the production capacity and the inventory and WIP dynamics. After adding the satisfaction rate constraints, the problem becomes a constrained optimization problem. We can either use the constrained OO that was introduced in Chapter V to solve this problem directly, or use a penalty function to convert the problem back to unconstrained optimization. Since we will demonstrate the application of COO in a remanufacturing system performance optimization problem in Section 3, we focus on the second way in this section. The penalty function is a quadratic function which is defined as follows,

$$\text{Penalty}\,(x; \beta) = \begin{cases} c(\beta - x)^2 & \text{if } x < \beta \\ 0 & \text{otherwise} \end{cases} \tag{8.3}$$

where the coefficient c is a penalty function. If c is very large, we will have a hard constraint, i.e., the selected design has to satisfy the constraint. The good enough set is defined as the top-n% designs. The parameter setting is almost the same as in experiment 1, except that the sales price, P_S, is $16, which is much lower. For the lower profit margin, we will keep a lower inventory level, and therefore the design that gives the optimum profit level will have a low satisfaction rate. With an interest in this problem, we will see the costs incurred when we enforce the high satisfaction rate constraint.

The results of the simulations are shown in Table 8.3.

Table 8.3. The results of the simulation when we have satisfaction rate constraints, where s is the number of designs selected by using the observed performance value (Note: Pre-determined upper bound for profit = $106,492)

s	J with no satisfaction rate constraint	J with $\beta = 0.97$	J with $\beta = 0.98$	J with $\beta = 0.99$
1	$96,030	$92,686	$93,819	$88,283
5	$96,413	$95,210	$93,819	$92,147
10	$96,413	$95,210	$94,022	$92,147
20	$96,413	$95,210	$94,022	$92,147
50	$96,413	$95,210	$94,022	$92,147

From the results in Table 8.3, we observed that if we have to enforce the satisfaction rate higher than 0.97, there will be a profit lost of $800[5]. This table, which is obtained within an hour, will be useful for a manager to know the cost associated with the satisfaction rate constraint. Actually, by using the OO-based method as a simulation-based optimization platform for the scheduling problems in apparel manufacturing systems, it is now possible to study many aspects of the system in a more quantitative way, such as the performance of new supply chain contracts between the manufacturers and the retailers (Bouhia 2004; Volpe 2005).

1.5 Conclusion

In this section, we apply the OO methods introduced in Chapter II to a scheduling problem in the apparel manufacturing system. We show how to incorporate the problem information in the initial random sampling of the designs and the construction of the crude model. The results are very promising. The OO-based method is very fast and only needs several

[5] When this constraint increases to 0.99, the cost incurred will be roughly $4,000.

minutes to screen out the good enough designs. We only use 1/2000 of the computation time that brute-force simulation would have taken in Experiment 1. The performance of the design found by OO is not only within the top-5% of the design space, but also within 3% from an upper bound of the optimum. This method supplies a simulation-based optimization platform to quantitatively analyze the performance of the apparel manufacturing system, which supplies many possibilities for further improving the performance of the apparel manufacturing system. Note that we only consider the linear model to generate the target level in this section. There are also other models to approximate the periodic property of the mean value of the demand better than the linear model. Interested readers may refer to (Lee 1997) for more details.

2 The turbine blade manufacturing process optimization problem

In this section, we consider a turbine blade manufacturing process optimization problem (Yang 1998). The integrated blade and rotor is manufactured via extrusion, which is similar to the manufacturing of plastic parts, but with much tougher high strength metal used and higher quality requirement on the product. As an optimization problem, such a manufacturing process is distinguished by the large number of parameter settings of all the operations and the difficulty to accurately evaluate the quality of the final production. On the one hand, the parameters, such as the initial size of the billet, the ram velocity of the plunger, and the ambient temperature of the work piece being processed, usually take continuous values. There are a huge and in principle infinite number of possible parameter settings combinations. On the other hand, for security and combat considerations, the aircraft usually has high quality requirements on the turbine blade. This quality depends on the physical property of the turbine blade, such as the effective strain field, the effective strain rate field, and the maximum load-stroke. These physical properties are determined by the deformation process of the work piece during the manufacturing, which can only be accurately described by the finite element method (FEM). It usually takes hours if not days to use FEM to simulate (calculate) the entire deformation process, and accurately evaluate the quality of the turbine blade thus produced. Giving the extremely large search space, with the lack of structure information (such as the gradient information) of the search space, it is computationally infeasible to find the optimal parameter settings using brute force. In this section, we show how OO can help to solve this problem. The FEM is a deterministic but complex calculation. Based on our previous discussion in Section II.6, OO can also be applied in this type of problem. One purpose of this section is to justify this by using a real-life example. By applying ordinal optimization, we are able to find a good enough parameter setting based on a computationally fast but crude model, and save the computing budgets by 95%, comparing with brute force. We formulate the problem in Section 2.1, show the application of OO in Section 2.2, and briefly conclude in Section 2.3.

2.1 Problem formulation

Peripheral blades and central rotor compose the primary parts of an airplane turbine engine. The quality and reliability of the turbine blade is important for the functionality of the aircraft engine. To meet the symmetry requirement of operation under high rotations, the blades must be balanced around the rotor, which is a difficult and costly production stage if the blade and the rotor are produced separately first and then fused together. To solve this problem, the integrally-bladed rotor (IBR) is invented, which, as the name shows, is a component that integrates the blade and the rotor manufacturing (Fig. 8.6). The high engine performance demands of customers require that these components be made from traditionally "difficult-to-process" materials such as high-temperature titanium alloys, intermetallics, and Nickel-based superalloys. Often these materials are stronger than conventional tool materials and require special tooling and high temperatures for processing. Coupled with this tooling constraint are the high-strength and high-reliability requirements, which call for good to excellent control of the final metallurgical structure and pedigree of the materials. These requirements lead to cautious designs of processing operations, often with redundant operations to ensure acceptable final metallurgical characteristics. Therefore, the manufacturing of such components consists of a significant part of the life-cycle cost of a turbine engine. With growing demands on aeronautical technology, there is a strong interest in the optimization of the manufacturing process of these components.

The manufacturing process of an IBR is shown in Fig. 8.7. First, the raw material is cast into the billet with the required radius and height. Then, hot isostatic pressing (HIP) is used to reduce the porosity of metals, which improves the mechanical properties and increases workability. After that, by hammering on the end, the billet is made shorter and thicker. This operation is called "upset". Due to the high quality requirement on the IBR, there are two forge stages. In the first stage, by heat treatment, the cylindrical billet melts down, flows into the blocker die, and is rammed into the expected shape. This is called the blocker forge. In the second, the work piece is further forged to a shape near the net shape of an IBR, thus called the near-net-shape (NNS) forge. After all these, the work piece is machined to IBR. Among these operations, the blocker forge is the most complex one, which involves various thermo-mechanical processes, so we focus specifically on the optimization of the blocker forging stage in this case study. Roughly speaking, there are several stages when filling the blocker die in the blocker forge. A more detailed discussion will be presented in Section 2.2 and illustrated by Fig. 8.9.

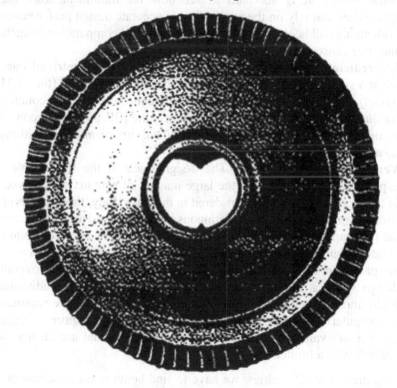

Fig. 8.6. An integrally-bladed rotor

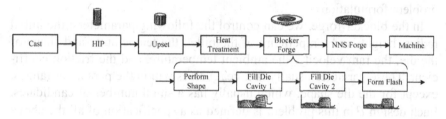

Fig. 8.7. The manufacturing process of an IBR

The three characteristics of the IBR problem that make it difficult to be solved by traditional optimization approaches can be described as follows.

1. **High complexity** in the structure of the problem. As Fig. 8.7 illustrates, many of the processes involved are nonlinear in nature and are interdependent. As a result, accurate evaluation of a single design (i.e., a specification of the parameter settings of the manufacturing process) via finite element calculations generally takes hours if not days or months to calculate, putting aside of the issue of a limited computational budget. It is not hard to see how the traditional searching approaches that rely on the availability of accurate design performance evaluations will become impractical under limited computation budgets and time constraints.

2. **Unpredictability of Inherent Imprecision.** In the nontrivial case when a simpler model that is different from the true model (the FEM model) is used for search, the model imprecision must be accounted for during the selection process. In essence, the IBR problem becomes a noisy search problem that most traditional optimization algorithms do not address.

3. **Very Large Search Space.** The design space of the IBR problem explodes exponentially with the large number of parameters involved. In the blocker forge stage considered in this case study, there are seven parameters (all defined over continuous interval ranges) and a choice of the die shape (from finite number of candidates) that can be controlled. (These parameters will be introduced later in this subsection.) Even if all parameters are discretized into 10 discrete values, the overall design space is roughly 10^7. With such a big space, the traditional build-and-test method is out of the question. For the same reasons, a computerized brute-force selection process will require a large number of evaluations, or numerical simulations, that are clearly in conflict with a limited time budget.

Giving the above difficulties, we have to find better ways to screen out some good parameter settings first, before we adopt the detailed FEM calculation. This is where OO helps. First, we introduce the mathematical problem formulation.

In the blocker forge, we can control the following parameters: the initial radius, height, and temperature of the billet, the temperature and shape of the die, the ram velocity, the ambient temperature, and the friction coefficient. These variables usually take real values within the parameter ranges, except for the die shape, which usually has a small number of candidates. Each design θ in this problem is defined as a specification of all the above seven parameter settings (the initial radius, height, and temperature of the billet, the temperature of the die, the ram velocity, the ambient temperature, and the friction coefficient) and a choice of the die shape. Then the design space Θ contains all the possible parameter settings. Given a design θ, through the detailed FEM simulation of the deformation process of the work piece during the operations, we can obtain the following physical quantities that determine the physical property of the IBR: the effective strain field, the effective strain rate field, the effective temperature field, the maximum stress field on the die surface, and the maximum load-stroke. The value of the cost function (definition follows) that evaluates θ, and the parameter setting of the operations, can be calculated.

The cost function used in this study consists of eleven terms, including five accounting costs, four quality loss penalties, and two inspection overheads. In the accounting cost category, the five cost factors considered are: material cost, initial reduction setup cost, initial reduction press cost, forge press cost, and die wear cost. The material cost refers to the market value of the initial billet. When the aspect ratio of the initial billet is too high, an initial reduction with a bottle-cap die set is necessary to avoid a buckling forge situation. The initial reduction setup cost and the initial reduction press cost refer to the setup cost and the operation cost per press run in this particular situation. The forge press cost is the cost of utilizing the press to complete the forge process. The die wear cost is the cost of the forge die set divided by the average number of production runs in the life of that TZM (Molybdenum Alloy) die set. The general equations[6] of the respective terms are:

Table 8.4. Accounting costs (Yang 1998)

Cost term	General form
Material cost	$C_{ma} \times$ total billet volume
Initial reduction setup cost	$C_{rs} \times \text{function}_{rs}$ (aspect ratio)
Initial reduction press cost	$C_{rp} \times \text{function}_{rp}$ (aspect ratio)
Forge press cost	C_{fp}
Die wear cost	$C_{dw} \times \text{function}_{dw}$ (billet temperature, maximum die pressure, processing time length)

In the penalty category, the four terms reflect the constraints on the material properties and the limits of the die capacity. These four terms are: force penalty, heat treatment penalty, heat remedy cost, and strain induced porosity (SIP) damage penalty. Force penalty reflects the maximum force

---

[6] Since our purpose here is to give a general picture of the complexity of the problem, we choose not to display the detailed mathematical formula. Actual detail can be found in (Yang 1998).

constraint of the press through a quadratic penalty function. The heat treatment penalty reflects the material strain constraint through an approximation of the fraction globalized in Ti64 material from the strain information of the work piece. The heat remedy cost specifies the cost of the heat treatment to remedy the strain imperfections in the work piece. The SIP damage penalty reflects the strain rate constraint and the temperature constraint on the final product through the estimation of the equilibrium volume fraction of the alpha phase for Ti64. The general equation, again in the spirit of footnote #1, for the four constraints are:

Table 8.5. Penalty terms (Yang 1998)

Penalty term	General form
Force penalty	$P_{fp} \times \text{function}_{fp}$(maximum die force)
Heat treatment penalty	$P_{ht} \times \text{function}_{ht}$(billet strain, billet temperature)
Heat remedy cost	$P_{hr} \times \text{function}_{hr}$(billet strain, billet temperature)
SIP damage penalty	$P_{SIP} \times \text{function}_{SIP}$(billet strain rate, billet temperature)

Finally, in the inspection category, the two terms are: forge setup inspection cost and ultrasonic inspection cost. The forge setup inspection is a fixed cost term to insure safety during the forge process. The ultrasonic inspection cost is the mandatory ultrasonic non-destructive evaluation prior to the acceptance of the final product. The general equation for two inspection overheads can be described as:

Table 8.6. Inspection overheads (Yang 1998)

Inspection overhead	General form
Forge setup inspection	O_{si}
Ultrasonic inspection	O_{ui}

The cost function can be summarized as follows:

$$
\begin{aligned}
J(\theta) = {} & C_{ma} \times \text{total billet volume } (\theta) \\
& + C_{rs} \times \text{function}_{rs}(\text{aspect ratio}(\theta)) \\
& + C_{rp} \times \text{function}_{rp}(\text{aspect ratio}(\theta)) \\
& + C_{fp} \\
& + C_{dw} \times \text{function}_{dw}(\text{billet temperature } (\theta), \\
& \qquad \text{maximum die pressure } (\theta), \\
& \qquad \text{processing time length } (\theta)) \\
& + P_{fp} \times \text{function}_{fp}(\text{maximum die force}(\theta)) \\
& + P_{ht} \times \text{function}_{ht}(\text{billet strain}(\theta))
\end{aligned}
$$

$$+ P_{hr} \times \text{function}_{hr}(\text{billet strain}(\theta))$$
$$+ P_{SIP} \times \text{function}_{SIP}(\text{billet strain rate}(\theta),$$
$$\text{billet temperature}(\theta))$$
$$+ O_{si}$$
$$+ O_{ui}.$$

2.2 Application of OO

The basic idea of OO is to use a crude model, which is computationally easy, to screen out quickly some good enough designs. From the last subsection, we can see that the value of the cost function depends on the parameter settings of the operations and the physical properties of the IBR thus produced. To accurately evaluate the physical properties of the IBR, the FEM model that describes the thermo-mechanical processes have to be used. In order to apply OO in this problem, it is crucial to find a crude model, which approximates the thermo-mechanical processes in a fast way and can give rough estimate of the physical properties of the IBR thus produced. Fortunately, the Ohio University Forge Simulation Model (the OU model) (Gunasekera et al. 1996) offers us such a choice. Compared with the FEM model, which contains all the details in the forging process, the OU model introduces the following simplifications. First, Gunasekera et al. showed that all changes in continuum properties such as strain, strain rate, and temperature can be described as functions of geometry or changes in the geometry with respect to time (Gunasekera et al. 1996). Based on this observation, instead of tracking down the changes in all the physical quantities at the same time like the FEM model does, the OU model only tracks down the change of the geometry of the work piece. Second, the general die shape (quarter cross-section view) is shown in Fig. 8.8. Instead of tracking the entire field of the continuum properties such as strain, strain rate, temperature, pressure, and grain size, the OU model divides the work piece into four parts: web, flange 1, flange 2, and flash. It calculates only the estimated average of these characteristic values in the regions with the assumption that these thermo-mechanical properties are uniform inside each region. Third, the evolution of the work piece during the forge process is simplified. When a billet is heated, input to the die, and rammed, it does not fill in every part of the die immediately. This process takes some time, and consists of five sub-stages as shown in Fig. 8.9. We can see that the shape of the work piece is not regular during these sub-stages. Instead of describing this deformation in details like the FEM model does, the OU model simplifies the five sub-stages as shown in Fig. 8.10. As we can see, the form of the work piece is more regular than in Fig. 8.9. Based on this simple approximation of the work piece geometry evolution, the OU model calculates the changes in the height and diameter, and calculates the strain and strain rate values. The calculation of other physical quantities, such as temperature, microstructure, die pressure, and the total die force, are calculated based on semi-empirical models together with some other approximations. In this way, the simulation is much simplified, and much faster than the FEM. A comparison study shows that it takes the FEM about 4 hours to evaluate one design, but only 0.1 seconds for the above crude model since it is made up of analytical formula (Yang 1998). This is a tremendous saving in computing time.

Note that the above crude model is a deterministic but simple calculation. Because the true model (i.e., the FEM) is too complex, the deterministic errors between the two models are complex and hard to predict. Based on our discussion in Section II.6, we can regard these errors as random noises, and treat the problem as if the true model is a stochastic simulation. We use a case study to justify these statements.

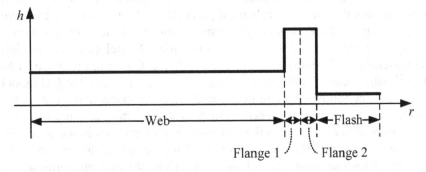

Fig. 8.8. General die shape (quarter cross-section view)

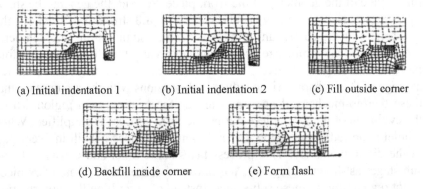

(a) Initial indentation 1   (b) Initial indentation 2   (c) Fill outside corner

(d) Backfill inside corner   (e) Form flash

Fig. 8.9. The geometry evolution of the work piece described by the FEM model (different time snaps of the work piece during the forge process) (Yang 1998)

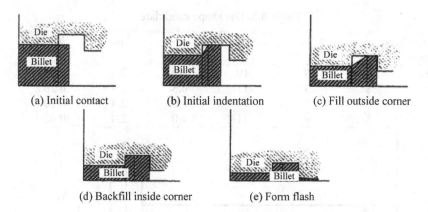

(a) Initial contact   (b) Initial indentation   (c) Fill outside corner

(d) Backfill inside corner   (e) Form flash

Fig. 8.10. The geometry evolution of the work piece described by the OU model (different time snaps of the work piece during the forge process)

In the following case study, we take the parameter ranges as shown in Table 8.7. The die shape can be defined by 6 variables, as shown in Fig. 8.11. There are four candidates for the die shape (Table 8.8). First we study the difference between the crude model and the true model (i.e., the FEM). We uniformly randomly sample 80 designs from the entire design space. We use the number 80 because it is too time-consuming to use the detailed model to accurately evaluate the performance for a large number of designs. Actually it takes about 14 days of continuous computing to finish the performance evaluation of these 80 designs using FEM. Comparing with so long a time, it is amazing how fast the crude model is. Only 8 seconds! We plot the observed cost vs. true cost in Fig. 8.12.

Table 8.7. Parameter ranges

Parameter	Parameter range ([min, max] unit)
Initial billet radius	[3, (flange radius – 0.5)] inch
Initial billet height	[(web height + 0.5), (3×initial billet radius)] inch
Die temperature	[1562, 1832] °F
Ram velocity	[0.1, 0.6] inch/sec
Initial billet temperature	[die temperature – 25, die temperature + 25] °F
Ambient temperature	[die temperature – 25, die temperature + 25] °F
Friction coefficient	[0.2, 0.8]

Table 8.8. Die shape candidates

Die shape index	r_1	r_2	r_3	h_1	h_2	h_3
0	8	9	10	1.0	2.4	0.15
1	8	9	11	0.5	2.4	0.20
2	8	9	10	1.0	2.4	0.20
3	8	9	11	1.0	2.4	0.05

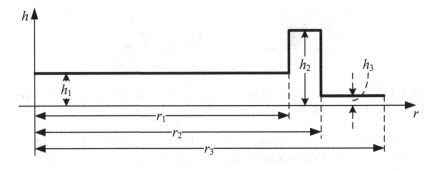

Fig. 8.11. Die shape variables

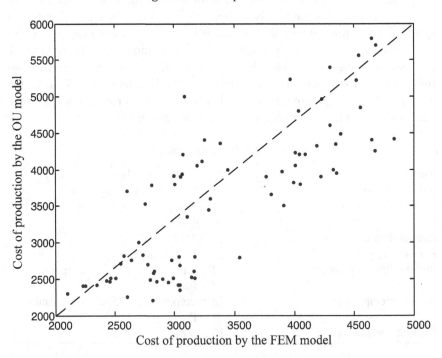

Fig. 8.12. Observed cost (by the OU model) vs. true cost (by the FEM model)

In Fig. 8.12, each dot on the graph represents a design with its x value being the true cost calculated by FEM, and its y value being the cost predicted by the crude model. The dotted line is the 45° line, on which all the points shall fall if the crude model conforms exactly to the FEM model. Data analysis shows that the prediction given by the crude model has an average %error (defined as (|crude model predicted value|-|FEM value|)/ |FEM value|) of 14% and a standard deviation of %error at 12%. The maximum %error observed is 62% and the minimum is 0%. This means the crude model does not give accurate performance evaluations. Actually in the 80-design instance shown in Fig. 8.12, the observed best design is the truly 20-th best. If we focus only on the best design, we can hardly succeed. Now, we apply OO to find some good enough designs with high probability.

Due to the extremely large computation needed to accurately evaluate 1000 designs (an estimate shows that it will take about 160 days to finish all the calculation), we only have the true performance of 80 designs. In the following, we will regard these 80 designs as the representative set Θ_N. Astute readers might notice that the UAP table in Section II.5 was obtained under the assumption of $N=1000$. They may ask whether it is reasonable to use that UAP table to estimate the size of the selected set when $N=80$. The numerical results, which will be shown later, justify this usage.

We show the observed OPC of these 80 designs in Fig. 8.13, which belongs to the neutral type. To see the difference between the OU model and the FEM model, we also show the corresponding true performances of these designs. Then we randomly select several of these 80 designs to estimate the normalized noise level, which is 0.1729. This belongs to the small noise level in the UAP table (Table 2.1 in Section II.5). For different values of the good enough set (g), the required alignment levels (k), the predicted values of s based on Eq. (2.42) are shown in Table 8.9, denoted as $\hat{s}_1$. Since the true noise level is smaller than 0.5, we use linear interpolation to obtain another group of predicted values of s, denoted as $\hat{s}_2$. This linear interpolation method will be explained in details later. For the instance of these 80 designs, we also present in Table 8.9 a size s^* whose value is decided such that there are at least k truly good enough (recall that we know the true performance of these 80 designs) designs in the observed top s^* ones. This quantity is shown here as a measure of the ideal size of the selected set to achieve the desired alignment level. We can see that $\hat{s}_1$ is always an upper bound of s^*, which shows the conservative nature of the UAP table. Now we show how to obtain less conservative estimate of s.

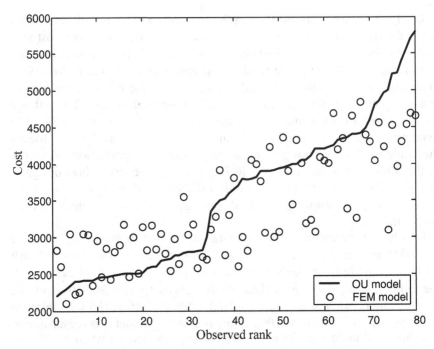

Fig. 8.13. The observed OPC

Table 8.9. The predicted and true selected sizes

g	k	s^*	$\hat{s}_1$	$\hat{s}_2$
1 (top 1.25%)	1	3	80	29
4 (top 5%)	1	3	13	6
	2	5	25	10
	3	6	37	15
	4	9	49	20
8 (top 10%)	1	3	6	3
	2	5	10	5
	3	6	15	8
	4	9	19	10
	5	11	24	12
	6	13	28	14
	7	17	33	16
	8	19	38	19

One important reason that $\hat{s}_1$ is conservative is that the true noise level is not 0.5, but 0.1729, which is much smaller. So we use linear interpolation to obtain better prediction of s. Note that if the noise level is 0, which means we know the true performance of all the designs. We only need to select the observed top-k designs (which are just the true top-k designs) to cover k truly top-g designs, i.e., $s = k$. Through previous calculation, we also know the predicted value of s when the noise level is 0.5, i.e., the $\hat{s}_1$ in Table 8.9. Through linear interpolation of size s in terms of the noise level, we obtain the prediction of s when the noise level is 0.1729 (denoted as $\hat{s}_2$ in Table 8.9). For example, when $g = 8$, $k = 1$, if the noise level is 0.5, $\hat{s}_1 = 6$; if the noise level is 0, s should be 1; now, the noise level is 0.1729, so the new estimate can be obtained from the following linear interpolation: $\lceil (6-1)/0.5 \times 0.1729 + 1 \rceil = 3$, which is denoted as $\hat{s}_2$ in Table 8.9. We can see that $\hat{s}_2$ is less conservative than $\hat{s}_1$, which usually is an upper bound of the true value s^*, and close to the true value s^* when $g = 8$. The only exception is when $g = 8$ and $k = 7$, the predicted $\hat{s}_2$ is smaller than s^*, but still very close. After goal softening, if we want to find at least one of the top-10% designs with high probability, the prediction $\hat{s}_2$ says we only need to investigate the observed top-3 designs. Comparing with brute force, we reduce the detailed performance evaluation by more than 25 folds (from 80 to 3). Through these numerical results, we see OO can help to save the computing budgets even when the objective function is not a stochastic simulation but a deterministic complex calculation. The results also justify the application of the UAP table when the representative set Θ_N is smaller than 1000.

2.3 Conclusion

In this section, we have considered a turbine blade manufacturing process optimization problem. The objective function can only be accurately evaluated through a complex but deterministic calculation. By using a crude model, which is more than 10000 times faster than the detailed FEM model, together with the idea of goal softening, we are able to save the computing budgets by more than 25 folds, comparing with brute-force calculation. This justifies that we can apply OO to solve the problem, taking the deterministic but complex error between the crude model and the detailed model as random noise. It should be noted that we omit many technical details to simplify the above discussion, such as the equations to describe the forging process in the FEM, and parameter settings of the 80 designs that are randomly sampled. Readers can refer to (Yang 1998) for more details. (Yang 1998) also shows that we can increase the accuracy of the cost prediction by incorporating more information in the crude model. This in turn helps to reduce the selected set thus required, and further save the computing budgets. We also show how to reduce the selected set size by interpolation. Another way to reduce the selected set size can be found in Section 3.3 below.

3 Performance optimization for a remanufacturing system

In this section we consider a remanufacturing system (Song et al. 2005a, 2005b). The goal is to manage the number of machines in the repair shop and the number of new parts to order in the inventory, so that the maintenance cost is minimized and the average maintenance time for an asset is not too long. Since there are two considerations in this problem, we can regard the maintenance cost as the objective function, and the requirement on the maintenance time as the constraint. In this way, we have a constrained optimization problem. The corresponding problem formulation will be discussed in Section 3.1 in details. Due to the time-consuming simulation-based evaluation of both the objective function and the constraint, we apply the constrained ordinal optimization as introduced in Chapter V. The application procedure is shown in Section 3.2. The experimental results are also presented, which is promising because we save the computing budgets by 25 folds. However, alternatively we can regard both the maintenance cost and the maintenance time as objective functions. We have then a two-objective function simulation-based optimization problem. We apply the vector ordinal optimization as introduced in Chapter IV. Especially, we show how to incorporate the problem information to further save the computing budgets in VOO. The details are presented in Section 3.3. We make a brief conclusion in Section 3.4.

3.1 Problem formulation of constrained optimization

Due to the consideration of saving the production cost and reducing environmental pollution, the study on remanufacturing system has attractted more and more interest recently (Guide et al. 1999; Guide 2000). The basic idea of remanufacturing system is to re-use the parts (sometimes after repair) from the old products to produce new products. This idea is especially useful for very expensive assets (such as aircraft jet engines) which consist of many parts. Rejecting old parts directly not only causes environmental pollution easily but also increases the production cost of a new asset. Thus the old parts are usually recycled after some repair.

A detailed model of a remanufacturing system is shown in Fig. 8.14. Due to random failure, the asset is shipped to this remanufacturing system. After disassembled into parts and inspected, the parts still in serviceable condition will be directly sent to a certain place and wait to be reassembled into new assets. The other parts need some repair and are sent to the repair shop. After the repair, the parts enter an inventory, and are then assembled into new assets together with the parts in serviceable condition, then leave the system. Since the parts of the same type are not distinguished from each other during the assembling, the inventory is also called the rotable inventory in practice (Kleijn and Dekker 1998). Due to the random arrival of the asset to this system and the uncertainties in the waiting time and repair time in the repair shop, sometimes there might not be enough parts when assembling a new asset. To avoid this "lack of synchronization", new parts can be ordered to the inventory. The parameters we can control are the number of machines in the repair shop and the number of new parts to order in the rotable inventory. We care about two performances metrics of the system. One is the average maintenance cost of an asset. The other one is the average maintenance (remanufacturing) time of an asset. These two performances are obviously related to one another. When there are more machines in the repair shop or more new parts are ordered in the inventory, the maintenance time can be reduced but the maintenance cost increases. The question is how we can minimize the maintenance cost[7], given the requirement on the maintenance time.

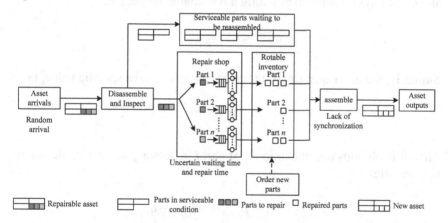

Fig. 8.14. Detailed model of the remanufacturing system

---

[7] Actually the opportunity to do preventive maintenance or not while the engine is disassembled is another decision which we will not consider in this example.

We mathematically formulate the problem as follows. Consider a planning horizon with m seasons (there are 3 months in each season). Suppose there are n parts in each asset, each of which requires a specific type of machine to repair. Let $C_{i,j}$ be the number of machines of type i that are used in season j. And let $\Delta I_{i,j}$ be the number of new parts ordered at the beginning of season j, which will be shipped to the inventory and become available in the next season. The maintenance cost consists of the machine cost ($\alpha \sum_{i=1}^{n} \sum_{j=1}^{m} C_{i,j}$), the cost for ordering new parts ($\sum_{i=1}^{n} \sum_{j=1}^{m} \Delta I_{i,j}$), and the inventory holding cost ($\beta \sum_{i=1}^{n} \sum_{j=1}^{m} I_{i,j}$), i.e.,

$$J(C,\Delta I) = \alpha \sum_{i=1}^{n} \sum_{j=1}^{m} C_{i,j} + \sum_{i=1}^{n} \sum_{j=1}^{m} \Delta I_{i,j} + \beta \sum_{i=1}^{n} \sum_{j=1}^{m} I_{i,j},$$

where α and β are positive real numbers, C and ΔI are n-by-m matrix, with $C_{i,j}$ and $\Delta I_{i,j}$ as the components. $C_{i,j}$ should not exceeds a specific value in two neighboring seasons, i.e.,

$$\left| C_{i,j-1} - C_{i,j} \right| \le \Delta C_i,$$

and $C_{i,j}$ should be controlled within a reasonable range, i.e.,

$$C_i^{\min} \le C_{i,j} \le C_i^{\max}.$$

Similarly, the order quantity of the parts cannot exceed a specific value, i.e.,

$$\Delta I_{i,j} \le \Delta I_i^{\max}.$$

Then it is obvious that the amount of part i in season j in the rotable inventory satisfies

$$I_{i,j} = I_{i,j-1} + p_{i,j-1} + \Delta I_{i,j-1} - q_{i,j-1},$$

where $p_{i,j-1}$ represents the number of part i that were finished in season j-1, $q_{i,j-1}$ represents the number of part i that were used in season j-1. Let $a(k)$ be the inter-arrival time between the k-th and the $(k-1)$-th asset. The first asset arrives at time $a(1)$, and the k-th asset arrives at time $\sum_{i=1}^{k} a(i)$.

After the k-th asset is disassembled and inspected, some parts are in serviceable condition and will be assembled with some other parts from the inventory into a new asset, then leave the system at time $\eta(k)$. The maintenance time for this asset is defined as

$$T(k) = \eta(k) - \sum_{i=1}^{k} a(i).$$

During the planning time horizon, the probability that the maintenance time exceeds a given limit T_D is

$$\text{Prob}\left[T(k) > T_D / \eta(k) \le T_C\right].$$

Suppose the constraint is that this probability should not be large, say less than P_0. Then the constrained optimization problem is

$$\min_{C, \Delta I} J(C, \Delta I) \text{ s.t. } \text{Prob}\left[T(k) > T_D / \eta(k) \le T_C\right] < P_0.$$

Although the above problem formulation might be simpler than the real system, this formulation preserves the basic characteristic of the real system, especially the difficulties. First, both the number of the machines and the number of new parts can only take discrete values. There are $n \times m$ variables in C and ΔI each, and thus the size of the design space is close to

$$\prod_{i=1}^{n} \left(\Delta I_i^{\max} \min \left\{ 2\Delta C_i, C_i^{\max} - C_i^{\min} \right\} \right)^m,$$

which grows exponentially as n and m increases. Simulation is the only way to do detailed performance evaluation for each $(C, \Delta I)$. Second, both the objective function and constraint are simulation-based. To obtain an accurate performance evaluation of the objective function and the constraint, we need ~1000 replications, which will take 30 minutes for each design by using the Enterprise Dynamics Software (Song et al. 2005a). If we want to accurately evaluate the feasibility of 1000 randomly sampled designs, we will need 500 hours, which is a very long time. In the next subsection, we apply constrained ordinal optimization to deal with these difficulties.

3.2 Application of COO

As introduced in Chapter V, the idea of COO is to use a feasibility model to quickly screen out the feasible designs (probably with some mistakes), and apply a crude model within these designs that are predicted as feasible to find some truly good enough and truly feasible designs. In order to apply COO to the remanufacturing system, we need to find an imperfect feasibility model for the constraint and a crude model for the performance.

3.2.1 Feasibility model for the constraint

By definition, any method that can predict the feasibility of a design with reasonable accuracy (say higher than 0.5) can be a feasibility model used in COO. This gives us a lot of freedom, such as heuristics and experiences. Of course, a feasibility model with higher accuracy will make a smaller number of mistakes, thus can further save the computing budgets. In this example, we use a machine learning method to obtain a feasibility model. The idea is as follows: First, we randomly sample a small number of designs, and then use brute-force simulation to accurately determine the feasibility of these designs. Input these designs and the corresponding feasibility as the training data, and use a machine learning method to discover the relationship between the parameter setting in the design and the feasibility. When the training finishes, we obtain a model. When a new design is input to this model, a predicted feasibility will be output. In this remanufacturing system, a feasibility model was found in this way. For technical details, such as what training method is used, and what the feasibility model looks like, please refer to (Song et al. 2005a). The feasibility model is very fast (0.003 second to predict the feasibility of a design on the average) and has a high accuracy, 0.985, which means if we randomly sample 1000 designs that are predicted as feasible by this feasibility model, an average of 985 designs are truly feasible. Although COO can work with a feasibility model with much less accuracy, such a high accuracy does allow us to save the computing budgets by 25 folds, as will be shown later in this subsection.

3.2.2 Crude model for the performance

After the designs predicted as feasible are screened out by using the feasibility model, we need a crude model to sort these designs according to the observed performance. The crude model should be computationally fast, and only need to give a rough estimate of the performance of the design. In one extreme case, blind pick does not need the estimate of the performance.

Since no problem information is utilized in the blind pick, the required size of the selected set is an upper bound of the case when other crude models are used, e.g., using a single replication of the simulation to estimate the performance.

The application procedure of the COO (feasibility model with blind pick) is summarized in Box 8.2.

Box 8.2. Application procedure of COO (feasibility model with blind pick) in the remanufacturing system

Step 1: Uniformly randomly sample N designs from the entire design space. Step 2: Use the feasibility model to screen out the predicted feasible designs. Step 3: User defines the size of the good enough set g and the required alignment level k. Step 4: Using the accuracy of the feasibility model, we can calculate the size of the selected set s. Step 5: Blind pick s designs from the predicted feasible list. Step 6: The COO theory ensures that there are at least k truly good enough and feasible designs in these s selected designs with high probability.

3.2.3 Numerical results

To get an idea of how much computing budget we can save by using COO, we show the following experimental results. Consider a planning for 8 seasons (24 months), $m = 8$, $T_C = 720$ days. Suppose there is one part that needs to be repaired (i.e., $n = 1$) after disassembly. The inter-arrival time in season i contains exponential distribution, i.e.,

$$\text{Prob}\left[a(k) = t\right] = \lambda_i e^{-\lambda_i t},$$

where the time unit is day, and the λ_i in the 8 seasons take the values of 3.5, 3.0, 2.5, 2.0, 2.5, 3.0, 3.5, and 3.0. It takes 5 days to disassemble and check each asset. The repairing time of the parts satisfies the triangular distribution, with a minimum of 30 days, a maximum of 90 days, and an average of 60 days. It takes 7 days to reassemble the parts. In this case, C and ΔI are both 8 dimensional row vectors, and the elements within are C_i and ΔI_i, representing the number of machines in season i, and the order quantity at the beginning of season i. $11 \leq C_j \leq 40$, $0 \leq \Delta I_j \leq 7$, $C^{max} = 40$, $C^{min} = 11$, $\Delta C = 5$, $\Delta I^{max} = 7$, $\alpha = 1$, $\beta = 0.2$, $T_D = 100$ days. The size of the design space is 5.8×10^{14}. The requirement on the maintenance time is:

$$\text{Prob}\left[T(k) > T_{\mathrm{D}}/\eta(k) \le T_{\mathrm{C}}\right] < 0.05,$$

i.e., P_0=0.05.

To get a rough idea on how many designs are feasible, we uniformly randomly sample 1000 designs and use brute force simulation to obtain the true performance and feasibility of these designs, as shown in Fig. 8.15. We can see that a lot of designs are not feasible. If we do not have a feasibility model and directly apply OO in this problem, there will be a lot of infeasible designs in the selected set, which leads to a large selected set.

We regard the truly top 50 feasible designs as good enough. For different alignment probability, we use Eq. (5.6) in Section V.1 to calculate the size of the selected set such that there is at least 1 truly good enough and feasible design in the selected set with a probability no less than the required alignment probability. These sizes are listed in Table 8.10.

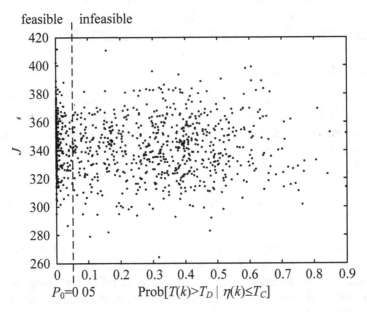

Fig. 8.15. The true performance and feasibility of 1000 randomly sampled designs

Table 8.10. Selected set size s for the remanufacturing system

Required AP	s
≥0.50	10
≥0.70	16
≥0.95	39

Suppose the required AP is 0.95. Since the designs are randomly sampled. The alignment level between the selected set and the good enough set might be different in different experiments. We show one instance in Table 8.11, where only the indexes of the designs are shown. In this instance there are 3 truly good enough and feasible designs found. Compared with the brute force simulation, which needs to accurately evaluate the performance and the feasibility of all the 1000 designs, COO saves the computing budgets by 25 folds in this example, by reducing from $N = 1000$ to $s = 39$.

Table 8.11. One instance of the alignment between G and S

Set	Plans
S	{404, 858, 744, 766, 245, *763*, 241, 466, 48, 532, 408, 906, 186, 39, 597, 577, 589, 351, 567, 406, 948, 882, 988, 402, 924, 464, 667, 530, 984, 906, 633, 357, *317*, 907, 119, 305, *857*, 737, 646}
G	{90, 270, 450, 630, 810, 990, 157, 337, 517, 697, 877, 1, 194, 374, 554, 734, 914, 136, 316, 496, 676, 856, 29, 209, 389, 569, 749, 929, 184, 364, 544, 724, 904, 43, 223, 403, 583, *763*, 943, 146, 326, 506, 686, 866, 137, *317*, 497, 677, *857*, 143}
$G \cap S$	{*317*, *763*, *857*}

3.3 Application of VOO

As aforementioned, in practice we sometimes do not know the appropriate value of P_0, which is the threshold for the probability that a maintenance time exceeds the given value. What we know is that the maintenance time is an important aspect of the system performance that should be considered during the optimization. We will here regard both the maintenance cost and the maintenance time as objective functions. More specifically we have two objective functions. One is the probability that the maintenance time exceeds a given limit, i.e.,

$$J_1(C, \Delta I) = \text{Prob}\left[T(k) > T_D / \eta(k) \le T_C\right].$$

The other one is still the maintenance cost, i.e.,

$$J_2\left(C,\Delta I\right)=\alpha\sum_{i=1}^{n}\sum_{j=1}^{m}C_{i,j}+\sum_{i=1}^{n}\sum_{j=1}^{m}\Delta I_{i,j}+\beta\sum_{i=1}^{n}\sum_{j=1}^{m}I_{i,j}\,.$$

Then we have a two-objective optimization problem

$$\min_{C,\Delta I}J\left(C,\Delta I\right)=\min_{C,\Delta I}\left(J_1\left(C,\Delta I\right),J_2\left(C,\Delta I\right)\right)^{\tau}.$$

Both objective functions can only be accurately evaluated by simulations. We will apply the vector ordinal optimization introduced in Chapter IV to solve this problem.

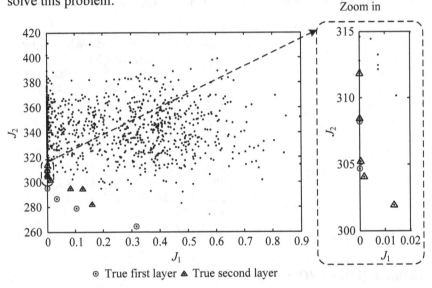

Fig. 8.16. The true performances of the designs

Since we have already used brute force simulation to obtain the true performance and feasibility of 1000 randomly sampled designs in Fig. 8.15, we show the first two layers of these 1000 designs in Fig. 8.16. There are 6 designs in the first layer (marked by circles), and 8 designs in the second layer (marked by triangles) shown in Fig. 8.16. We also show the observed performance curve in the vector case (VOPC) in Fig. 8.17. Of course, in practice we do not know this true VOPC. Instead, we use a crude model (a single replication) to get the rough estimation of the performance. Fig. 8.18 shows one instance, where there are 4 designs in the observed first and the observed second layer, respectively, which is different from Fig. 8.16. The VOPC is similar to Fig. 8.17, which belongs to the steep type. We estimate the noise level by 10 independent simulations of a design, and find the normalized noise level is 0.1061 for J_1 and 0.0078 for J_2, which is a small noise level. By looking at the VOO-UAP table Table 4.1 in Chapter IV, we find the coefficients in the regression function are: $Z_1 = -0.7564$, $Z_2 = 0.9156$, $Z_3 = -0.8748$, and $Z_4 = 0.6250$. Define the designs in the truly first two layers as good enough designs (there are 14 design in total). Because the VOO-UAP table is developed for 10,000 designs and 100 layers in all,

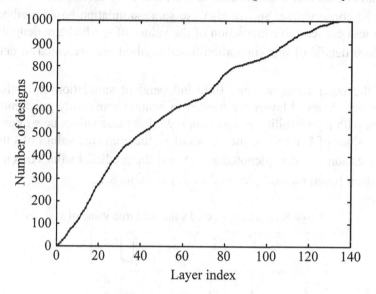

Fig. 8.17. The true VOPC

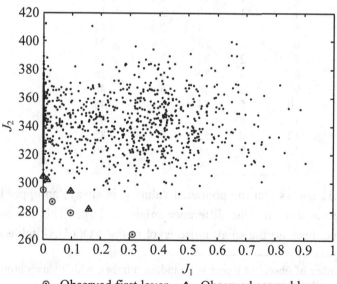

⊙ Observed first layer ▲ Observed second layer

Fig. 8.18. The observed performance of the remanufacturing system

we need to normalize the values of g, s, and k before the calculation. There are 134 layers in Fig. 8.17[8], so $g'=[100/134\times2]=1$, $k'=10000/1000\times k=10k$ ($1\leq k\leq14$), and $s=[134/100\times s']$. For different values of k, we denote the values of s predicted in this way as $\hat{s}_1$. Since the noise levels in J_1 and J_2 are smaller than 0.5, which is used in the VOO-UAP table, the values of $\hat{s}_1$ might be conservative. So, we also use some simulation-based method to obtain less conservative prediction of the values of s, which are denoted as $\hat{s}_2$. More details of this simulation-based method are presented in details later.

On the other hand, we use 1000 independent simulations to estimate how many observed layers are needed to contain some truly good enough designs with a probability no less than 0.95. Let these values be s^*. For different values of k, we show the values of s^*, the predicted values of s using the regression function (denoted as $\hat{s}_1$), and the predicted values of s using simulation-based method (denoted as $\hat{s}_2$) in Table 8.12.

Table 8.12. The predicted value and true value of s

k	s^*	$\hat{s}_1$	$\hat{s}_2$	$\left| \bigcup_{i=1}^{\hat{s}_2} \mathcal{L}_i \right|$
1	1	7	1	6
2	1	11	1	6
3	1	15	2	14
4	1	20	2	14
5	2	24	2	14
6	2	28	3	24
7	2	32	3	24
8	3	36	4	36
9	3	40	5	48
10	5	44	5	48
11	6	48	6	62
12	8	52	8	85
13	9	56	9	96
14	13	59	12	141

Table 8.12 shows that the predicted value $\hat{s}_1$ is always an upper bound of the true value s^*, but the difference might be large. This means $\hat{s}_1$ is conservative, because the small noise level in the VOO-UAP table is 0.5,

---

[8] The number of observed layers is a random number, which varies around the true number of layers.

but the noise level in this problem is 0.1061 and 0.0078, respectively. To obtain a less conservative estimate of s^*, we use the following method. Based on the rough estimate obtained through a single simulation, regard the estimate as the true value, find the observed first two layers, and regard these designs as the "truly good enough designs". Then add the normally distributed noise $N(0, 0.2122^2)$ and $N(0, 0.0156^2)$ [9], and find the observed first s layers. Repeat this procedure for 1000 times, and obtain an estimate of the alignment probability for each (s, k) (in which $s = 1, 2, \ldots 20$, and $k = 1, 2, \ldots, 14$). Select the smallest s such that the alignment probability is no less than 0.95 as the estimate of the true values s^*, and denote as $\hat{s}_2$, also shown in Table 8.12. We can see that $\hat{s}_2$ is a good estimate of s^*, and a tight upper bound of s^*, except for the case of $k = 14$. In the case of $k = 14$, the difference is only 1. From Table 8.12, we can see that VOO saves the computing budgets by at least one order of magnitude in most cases. Only when $k = 14$, the number of designs in the observed first $\hat{s}_2$ layers increases drastically. That is because it is a difficult job to find *all* the designs in the truly first two layers.

To summarize, the application procedure of VOO in this remanufacturing system is as follows.

Box 8.3. Application procedure of VOO in the remanufacturing system

Step 1: Randomly sample N designs.
Step 2: Use the crude model to quickly estimate the performance of the designs.
Step 3: Layer down the designs according to the observed performance.
Step 4: User defines the good enough set and the required alignment level k.
Step 5: Estimate the observation noise level and the problem type (VOPC).
Step 6: Use the VOO-UAP table in Section IV.2 to calculate the number of observed layers to select.
Step 7: Select the observed first s layers as the selected set.
Step 8: Then the VOO theory ensures that there are at least k truly good enough designs in the selected set with high probability.

---

[9] The standard deviations of the additive noises are twice the standard deviations of the noises in the two objective functions, respectively.

3.4 Conclusion

In this section, we consider the performance optimization of a remanufacturing system. By formulating the requirement on the maintenance time as a constraint or an objective function, we have a simulation-based constrained optimization problem or a two-objective optimization problem. We apply COO and VOO to solve the problems, respectively. Besides the general application procedure of COO and VOO as introduced in Chapter V and IV, we also discuss how we can incorporate the problem information into a feasibility model, using a machine learning method in COO; and how we can use problem information to improve the estimate of the number of observed layers to select in VOO. In both problems, COO and VOO save a lot of computing budgets. Note that we only use blind pick in Section 3.3. By using horse race selection rule, we can further improve the performance of COO by selecting a smaller number of predicted feasible designs. Interested readers may refer to (Song et al. 2005b) for more details.

   Exercise 8.1: Can we apply COO to solve simulation-based multi-objective optimization problems? If so, please explain how. If not, please explain why.

   Exercise 8.2: Can we apply VOO to solve simulation-based constrained optimization problems? If so, please explain how. If not, please explain why.

4 Witsenhausen problem

A celebrated problem in system and control is the so-called Witsenhausen problem. In 1968, H. S. Witsenhausen (Witsenhausen 1968) posed an innocent looking problem of the simplest kind. It consists of a scalar linear dynamic discrete time system of two time stages (thus involving two decisions at time stages one and two). The first decision is to be made at time one with perfect knowledge of the state, and there is a quadratic cost associated with the decision variable. The second decision can only be made based on noisy Gaussian observation of the state at time stage two, however, there is no cost associated with the decision. The performance criterion is to minimize the quadratic terminal state after the two decisions. Thus, it represents the simplest possible Linear-Quadratic-Gaussian (LQG) control problem except for one small detail: *Instead of the usual assumption of one centralized decision maker who remembers at time stage two what s/he knows at time stage one, we do not have perfect memory or recall.* In fact, we have a decentralized team problem with two decision makers (DMs), DM1 and DM2 who do not have complete knowledge of what the other knows. Here the possibility for optimization is clear. DM1 knows the state of the system perfectly. S/he can simply use his/her control variable to cancel the state perfectly and leave DM2 nothing more to do. However, his/her action entails a cost. On the other hand, DM2 has no cost to act, but, without perfect memory, s/he has no perfect knowledge of the state of the dynamic systems at time stage two. A simple approach would be to strike a compromise using linear feedback control law for each decision maker, which is also known to be optimal under the traditional centralized LQG system theory for problems with perfect memory. In fact, it is easy to prove that such a solution is a person-by-person optimal solution in equilibrium, i.e., if DM1 fixes his/her linear feedback control law, the best response by DM2 is a linear feedback control law and vice versa. However, Witsenhausen demonstrated that, without perfect memory, there exists a nonlinear control law for both DM1 and DM2, which involves signaling by DM1 to DM2, using its control action (The idea of signaling will be explained in more details in Section 4.1.), that outperforms the linear person-by-person optimal control law. In other words, the Witsenhausen problem presents a remarkable counterexample which shows that the optimal control law of LQG problems may not always be linear when there is imperfect memory. At the time, this was totally surprising since the problem seemed to possess all the right mathematical assumptions to permit an easy optimal solution. However, the globally optimal control law for such a simple LQG problem (or team decision problem) was unknown. The discrete version of the problem was known to be NP-complete (Papadimitriou and Tsitsiklis 1986). Many attempts and papers on the problem followed in the next thirty and more years before the problem was understood and a numerical solution of the globally optimal control law obtained in (Lee et al. 2001).

The difficulty of the problem constitutes the essence of information structure (who knows what and when) in decentralized control, which is a subject worthy of a separate book. We shall not go into the matters here. However, we shall use the Witsenhausen problem here to illustrate the process of search in the space of control laws, using OO to get good enough solutions. This is because there are so much data accumulated with this problem and we can easily assess the "good enough"-ness of any results thus obtained via OO.

First, we show the mathematical problem formulation in Section 4.1. Since the optimal control laws associated with the Witsenhausen problem have not been obtained analytically yet, it is important to discover the structure of the space of the control laws numerically. This is a common problem faced by many practical engineering problems, where finding the optimal design needs tremendous computing time, even if not computationally infeasible. By using OO, we were able to discover some structure information of the design space in the Witsenhausen problem efficiently, based on only noisy performance observation. The information not only produced a pair of control laws that were 47% better than the best solution known by that time (Banal and Basar 1987), but also helps to finally achieve the best-so-far numerical solution (Lee et al. 2001). We introduce the details for this also in Section 4.1. In Section 4.2, we consider the constraint on memory space when solving the Witsenhausen problem, and show how to find simple and good enough control laws using OO and OBDD, which were introduced in Chapter VI. With minor performance degradation (less than 5%), we save the memory space to store the control law by over 30 folds. We make a brief conclusion in Section 4.3.

4.1 Application of OO to find a good enough control law

The Witsenhausen problem can be described as follows. It is a two-stage decision making problem. At stage 1, we observe the initial state of the system x. Then we have to choose a control $u_1 = \gamma_1(x)$ and the new state will be determined as $x_1 = x + u_1 = x + \gamma_1(x)$. At stage 2, we cannot observe x_1 directly. Instead, we can only observe $y = x_1 + v$, where v is the additive noise. Then we have to choose a control $u_2 = \gamma_2(y)$ and the system state stops at $x_2 = x_1 + u_2$. The cost function is $E[k^2(u_1)^2 + (x_2)^2]$ with $k^2 > 0$ as a constant. The problem is to find a pair of control functions (γ_1, γ_2) which minimize the cost function. The trade off is between the costly control of γ_1 which has perfect information and the costless control γ_2 which has noisy information. We consider the famous benchmark case when $x \sim N(0, \sigma^2)$ and $v \sim N(0,1)$ with $\sigma = 5$ and $k = 0.2$.

Witsenhausen made a transformation from (γ_1, γ_2) to (f, g), where $f(x) = x + \gamma_1(x)$ and $g(y) = \gamma_2(y)$. Then the problem is to find a pair of functions (f, g) to minimize $J(f, g)$ where

$$ J(f,g) = E\left[k^2 \left(f(x) - x \right)^2 + \left(f(x) - g(f(x) + v) \right)^2 \right]. \qquad (8.4) $$

The first term in Eq. (8.4), $E[k^2(f(x)-x)^2]$, represents the cost shouldered by player one in the first time stage, so it is also called the stage one cost. The second term, $E[(f(x)-g(f(x)+v))^2]$, represents the cost shouldered by player two in the second time stage, so it is also called the stage two cost. Witsenhausen (Witsenhausen 1968) proved that: 1) For any $k^2 > 0$, the problem has an optimal solution. 2) For any $k^2 < 0.25$ and $\sigma = k^{-1}$, the optimal solution in linear control class with $f(x) = \lambda x$ and $g(y) = \mu y$ has $J^*_{\text{linear}} = 1 - k^2$, and $\lambda = \mu = 0.5\left(1 + \sqrt{1 - 4k^2}\right)$. In the benchmark case that we consider, $k = 0.2$, $J^*_{\text{linear}} = 0.96$. 3) There exist k and σ such that J^*, the optimal cost, is less than J^*_{linear}, the optimal cost achievable in the class of linear controls. Witsenhausen gave the following example. Consider the design: $f_W(x) = \sigma\, \text{sgn}(x)$, $g_W(y) = \sigma \tanh(\sigma y)$, where sgn($\bullet$) is the sign function, then the cost function J is $J_W = 0.4042$. 4) For given $f(x)$ satisfying $E[f(x)] = 0$ and var$[f(x)] \leq 4\sigma^2$, which are the conditions that the optimal $f^*(x)$ should satisfy, the optimal g^*_f associated with function f is

$$g^*_f = \frac{E\left[f(x)\varphi\left(y - f(x)\right)\right]}{E\left[\varphi\left(y - f(x)\right)\right]}, \tag{8.5}$$

where $\varphi(\bullet)$ is the standard Gaussian density function.

Now the problem becomes that of searching for a single function f to minimize $J\left(f, g^*_f\right)$. Although the problem looks simple, no analytical method is available yet to determine the optimal f^*. The numerical optimal solution only came after over thirty years later and after many attempts (Lee et al. 2001). In the following, we will demonstrate how we should apply OO to search for good control laws for the Witsenhausen problem. Before we present the numerical details, we should discuss what a crude model is and how we can apply OO to discover the property of the good enough designs, which helps to narrow down the search space.

4.1.1 Crude model

Following the properties of the optimal control laws shown by Witsenhausen, each "design" in the Witsenhausen problem is a control function f, which satisfies $E[f(x)] = 0$ and var$[f(x)] \leq 4\sigma^2$. Because f is in general a one-dimension real function, and there are in principle infinite number of such functions, it is important to find an appropriate representation for such functions in a digital computer. The idea is to discretize the function f. Lee et al. showed that it is reasonable to assume the optimal function f^* is symmetric about the origin, i.e., $\gamma_1(y_1) = -\gamma_1(-y_1)$ (Lee et al. 2001). In the following discussion, we only consider $f(x)$ for $x \geq 0$. We divide the x-space

[0,∞) evenly in probability, i.e., we divide x-space into n intervals, $I_1,...,I_n$, where $I_i=[\sigma t_{(0.5+0.5(i-1)/n)}, \sigma t_{(0.5+0.5i/n)})$, t_α is defined by $\Phi(t_\alpha)=\alpha$ where Φ is the standard normal distribution function. Prob$[x\in I_i]=0.5/n$ because x has a normal distribution $N(0,\sigma^2)$. Then for each interval I_i, a control value f_i is uniformly picked from $(-3\sigma, 3\sigma)$, i.e., $f_i\sim U(-15, 15)$. To calculate the performance $J\left(f,g_f^*\right)$, we should calculate the optimal associated function g_f^* through Eq. (8.5) and then we will obtain $J\left(f,g_f^*\right)$ through Eq. (8.4). However, both Eq. (8.4) and Eq. (8.5) involve expectations, which mean a large number of Monte Carlo simulations might be required to calculate the performance J accurately. Actually based on the results described in this subsection, (Lee et al. 2001) developed a step-function representation of the function f and then obtained a way to calculate J through numerical integration instead of Monte Carlo simulation. Although numerical integration is much faster than Monte Carlo simulation, it requires a long-time numerical integration to make the result very accurate. We will come back to this at the end of this subsection. Right now, let us assume Monte Carlo simulation is the only way to accurately evaluate J, which is true when the results of this subsection were developed in 1999.

The question now is how to find a crude model which is computationally fast and can give a rough estimate of J. We simplify the calculation from two aspects. First, instead of calculating the accurate g_f^*, we calculate an approximation of g_f^*,

$$\hat{g}_f = \frac{\sum_{i=1}^{100} f(x_i)\varphi(y-f(x_i))}{\sum_{i=1}^{100} \varphi(y-f(x_i))}.$$

Second, instead of using a large number of Monte Carlo simulations to accurately calculate $J\left(f,\hat{g}_f\right)$ through Eq. (8.4), we use only 100 replications to get an estimate $\hat{J}\left(f,\hat{g}_f\right)$. In this way, we get a crude model $\hat{J}\left(f,\hat{g}_f\right)$ of the true performance $J\left(f,g_f^*\right)$.

4.1.2 Selection of promising subsets

After the discretization of function f, the design space in the Witsenhausen problem is still extremely large. To overcome this difficulty, our basic idea is to divide the entire design space into smaller subsets and choose promising subsets for further searching. The No-Free-Lunch Theorem tells us that every other optimization method can be as efficient (inefficient) as blind pick, without any problem information. To achieve a higher efficiency than that of blink pick, we need to discover some structure information of the design space, e.g., which subset contains more good enough designs than others. In particular, if we sample a set of designs for their performances (however noisily or approximately), we should be able to catch a glimpse, from the samples, of what are "good" subsets to search, and gradually restrict the search there. This is like traditional hill climbing, except that we move from one subset to another instead of moving from point to point in the search space. The key question here is to establish a procedure of comparing two subsets based upon sampling and then to narrow down the search space step by step. We will show how OO helps to do this comparison in this subsection. By the OO-based comparison, we find three restrictions which narrow down the search space to a subset that contains more good enough designs than without the restrictions. The three restrictions are: 1) For each interval I_i, control f is in $(-0.5\sigma, 2.5\sigma)$ because we are searching primarily in the positive quadrant; 2) f is a non-decreasing function; 3) f has two steps. In the rest of this subsection, we will use numerical results to show how these three restrictions are discovered, and how this finally helps us to find a control law 47% better than the best solution known by that time, and later on helps to discover the best-so-far control law. First, we start from a general discussion on how to narrow down the search space using OO.

The idea behind narrowing down the search space is to identify which subset of the search space contains more good enough designs than the others. Of course, if we know the true performance of all the designs, this problem will be trivial. However, in practice we only have noisy performance observations and can only do this performance estimation for a small portion of the entire design space. So we need a function to represent the goodness of a subset, i.e., the number of good enough designs in this subset, and we hope this function can be efficiently estimated when only the noisy performance observations of some of the designs are available. The performance distribution function satisfies these constraints. Suppose Θ_1 and Θ_2 are two subsets of a large design space Θ. If we know the true performance of all the designs in Θ_1 and Θ_2, we can obtain a performance density function (Fig. 8.19(a)), which looks like a probability density function, by discretizing the performance into many small intervals, counting the number of designs the performances of which fall in each interval, and normalizing this number by the total number of designs. By integrating the performance density function, we obtain the performance distribution function (PDF) (Fig. 8.19(b)), which is non-decreasing and looks like a probability distribution function. Now, suppose we know the PDF of Θ_1 is $F_1(t)$ and the PDF of Θ_2 is $F_2(t)$, Let us focus on the top-5% designs in each subset. If $F_1(t_1)=0.05$, $F_2(t_2)=0.05$, and $t_1<t_2$, (as shown in Fig. 8.20) which means the top-5% designs in Θ_1 are with better performances than those in Θ_2. Then we should continue our search in Θ_1. Of course, in practice we only have the estimate of F_1 and F_2. As we will demonstrate by the numerical examples in the following, however, the estimate of F_1 and F_2 still helps us to find the promising subset.

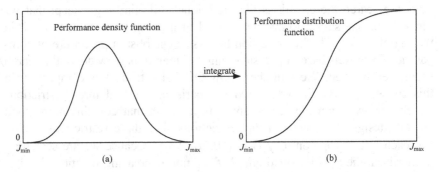

Fig. 8.19. (a) Performance density function and (b) performance distribution function

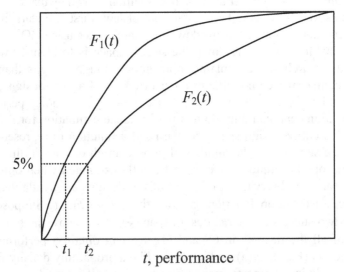

Fig. 8.20. Comparing two subsets based on PDFs

Now we are ready to summarize our sampling and space-narrowing procedure. For a search space Θ, we first define two or more subsets (there might be intersection among these subsets) and find the corresponding observed PDFs. By comparing the observed PDFs, we can estimate which subset(s) is (or are) good. We can then further narrow down our search into smaller subsets. In the following, we demonstrate how this procedure helps to discover the three properties of the good designs in the Witsenhausen problem and finally a design which was the state of the art when the result was published in year 1999 (Deng and Ho 1999).

The number of intervals, n, used in the construction of the discretized f, determines the size of the search space. In principle, n can be any positive integer. However, for a large n, the designs with good performances are only a very small portion of the entire search space. In this case, it is very difficult to find a good enough design. The first question here is what the appropriate value of n should be. For values of $n=1, 2, 5,$ and 10, we randomly pick 5000 functions f in each case, use the crude model in Section 4.1 to estimate the performances, and observe the PDF respectively (shown in Fig. 8.21). Since we are doing minimization, we only care about the performances of the top-5% designs in each case. In Fig. 8.21, we only show the part of the PDFs that are close to the origin. There are 4 curves, from the left to the right, representing the PDFs of the subsets when $n=1$, 2, 5, and 10, respectively. We compare the performance of the top-5%-th design in each subset and find that the cases of $n=5$ and 10 are of much larger costs (both greater than 1.0) than those of $n=1$ and 2 (both smaller than 1.0). Thus, $n=1$ and 2 are better subsets than $n=5$ and 10. The problem is the cases of $n=1$ and 2 are indistinguishable. Because if we only care about top-5% designs in both cases, many top-5% designs in the case of $n=2$ are with larger costs than the case of $n=1$. However, if we consider the top-0.1% designs, the case of $n=2$ is better than $n=1$. Through this comparison, we find that $n=1$ or 2 are good choices, but we cannot determine which one is better yet. The fact that $n=1$ is a good choice indicates that the class of constant control functions (with discontinuity at the origin due to symmetry) is a good representation of the search space for control function f. This means that the pair of controllers described by Witsenhausen which outperform the optimal linear control law are already very good. In addition, by comparing the observed best f among the randomly sampled 5000 functions f for each n, we observe that the control values of the observed best controllers for different n are located in $[-2, 12)$. Based on this observation, we make the following restriction (R1): *For each interval I_i, control f is in $(-0.5\sigma, 2.5\sigma)$, i.e., f~$U(-2.5,12.5)$.*

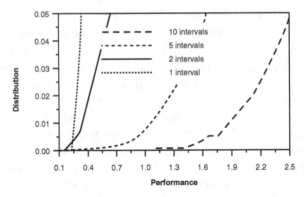

Fig. 8.21. Observed PDFs when there are $n=1$, 2, 5 and 10 intervals in $f(x)$, $x\geq0$ (Deng and Ho 1999) © 1999 Elsevier

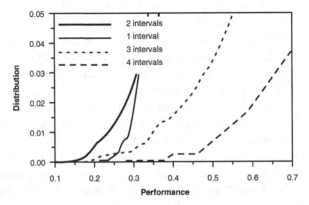

Fig. 8.22. Observed PDFs with restriction R1 (Deng and Ho 1999) © 1999 Elsevier

With this restriction, we repeat our experiment for $n=1$, 2, 5, and 10 and find the cases of $n=1$ and 2 still outperform the cases of $n=5$ and 10. We then do a further comparison among the cases of $n=1$, 2, 3, and 4. We show the observed PDFs and the observed best-control functions in Fig. 8.22 and Fig. 8.23. In Fig. 8.22, the top-5%-th design in the cases of $n=1$ and 2 are of costs smaller than 0.4, which is smaller than the costs of the top-5%-th designs in the cases of $n = 3$ and 4 (with costs larger than 0.5). If we compare the performances of the top-3%-th designs in each case, the case of $n = 2$ will be the best subset among the four cases. A more interesting phenomenon we may observe from Fig. 8.23 is that the observed best controllers for both $n = 3$ and 4 have the two-interval shape as the one of $n = 2$. All these observations indicate that the right direction of search should be toward the two-interval functions. Since the observed best controllers (in Fig. 8.23) display some increasing property, we make a further restriction (R2): *The control f is a nondecreasing function in $(-0.5\sigma, 2.5\sigma)$.*

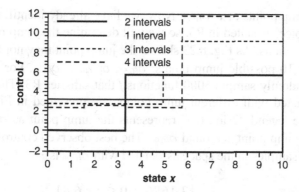

Fig. 8.23. Observed best control *f* with restriction R1 (Deng and Ho 1999) © 1999 Elsevier

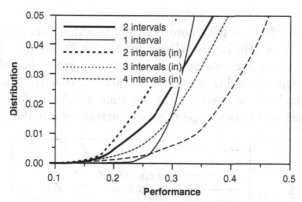

Fig. 8.24. Observed PDFs with restriction R2 (Deng and Ho 1999) © 1999 Elsevier

To test whether restriction R2 helps us to find subsets containing more good designs, we compare the observed PDFs before and after the restriction R2 is applied in Fig. 8.24. The curves with the legend "interval(in)" are those with restriction R2. Fig. 8.24 shows that with restriction R2, the top-5% designs in the two-interval controllers have the best performances. This indicates that the specification of the non-decreasing control function is in the right direction. Actually it was shown by Witsenhausen that the optimal function *f* should be non-decreasing (Witsenhausen 1968). It is conceivable that the optimal control function may possess significant discontinuity. Thus, the 3rd restriction (R3) is made as follows: *The control f is a two-value non-decreasing step function in (–0.5σ, 2.5σ).*

In the previous experiments, to make a quick estimate of the PDF for each value of *n*, when the number of intervals n is given, we fix the discretization of the *x*-space as explained in Section 4.1. For example, for *n*=2, we fix the jump points at $x=\sigma t_{0.75}$. Now, we have already identified that $n=2$ is a good choice, as stated in R3, so we will determine the jump point of the two-value functions. As Fig. 8.23 shows, the jump point may not be at $\sigma t_{0.75}$. We consider 10 possible jump points: $\sigma t_{0.55}, \sigma t_{0.60}, \ldots \sigma t_{0.95}$. For each jump point, we randomly sample 5000 functions f that satisfies R3. The observed PDFs, associated with different jump points, are presented in Fig. 8.25. In Fig. 8.25, the legend "2 int. (a)" represents the jump point as σt_a. We see that the best jump point is around $\sigma t_{0.90}$. The best observed control function f among 5000 samples in the space associated with $\sigma t_{0.90}$ is

$$f_{DH}(x) = \begin{cases} 3.1686, & 0 \le x < 6.41, \\ 9.0479, & x \ge 6.41. \end{cases}$$

The subscript "DH" is to denote that this function was first found by M. Deng and Y.-C. Ho in 1999 (Deng and Ho 1999). We use 10000 replications to obtain an accurate estimate of the true performance of this function, and obtain the value 0.1901 with variance 0.0001, which is 47% better than the best solution know by that time which was found by Banal and Basar with performance $J_{BB}=0.3634$ (Banal and Basar 1987).

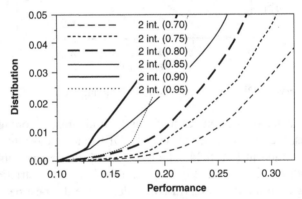

Fig. 8.25. Observed PDFs with restriction R3 (Deng and Ho 1999) © 1999 Elsevier

The after-the-fact reasoning behind the superiority of the two-value controllers is as follows. Witsenhausen proposed the signaling concept when he reported this famous counter example in (Witsenhausen 1968). The idea is that DM1 knows the state of the system perfectly but has an action cost. So instead of using his/her own control variable to cancel the state perfectly, DM1 cancels (or enhances) parts of the state (x), which makes the state of the system concentrated on either a given negative or positive point x_1. DM1 uses x_1 as a signal to tell DM2 how to set his/her control variables, such as positive or negative values of x_1. Under moderate noise conditions, DM2 can ascertain the sign of x_1 with high probability. DM2 has no action cost. If DM2 can interpret DM1's signal x_1 correctly, and thus takes the correct action to cancel almost all the state x_1, the resulting state of the system, x_2, will cause little cost, i.e., a small stage 2 cost. However, DM2 only has noisy observation of x_1, and may misinterpret x_1. To reduce the probability of misinterpretation, DM1 needs to take large actions to make the signaling levels far apart from each other, which causes a large stage 1 cost. Finding the optimal $f^*(x)$ in the class of step functions (as used in this section) amounts to finding the optimal number of steps/intervals in $f(x)$ and their placements so as to balance the tradeoffs between the first and second stage costs. The step function f_W proposed by Witsenhausen is a one-interval function (the single jump point is at $x=0$ and with signaling level σ). Banal and Basar further optimized the signaling level of this one-interval function and obtained f_{BB} (the single jump point is still at $x=0$ but with signaling level $\sigma\sqrt{2/\pi}$). The signaling scheme in f_{DH} allows DM1 to use four signal levels, i.e., more positive (9.0479), less positive (3.1686), less negative (−3.1686), and more negative (−9.0479). Hence, there is a reduction in the magnitude of $(x-f_{DH}(x))$. Meanwhile, the signaling levels are placed sufficiently far apart so that DM2 can still distinguish DM1's signal with small errors.

In the above discussion, we continuously narrow down the search space by comparing the observed PDFs of different subsets. The ideas of goal softening and ordinal comparison allow us to discover some properties of the good designs, i.e., the three restrictions we found. This finally led us to a design 47% better than the best solution known by that time (Banal and Basar 1987) when this result was published in 1999. These results indicate that a step function may be an appropriate representation of the function f. This idea was further explored in (Lee et al. 2001). In (Lee et al. 2001), by describing the function f as a step function, Lee et al. achieved a fast and accurate computational scheme for the cost J which eliminates the need of simulation, but requires numerical integration. Also they observed that the jump points should be located around the average of two adjacent values of function f (which is also called the signaling levels). Furthermore, additional improvement can be made by adding small segments to approximate a slight slope for each step in function f. Finally they achieved the following function f

$$f_{\text{LLH}}(x) = \begin{cases} 0.00 & 0.00 \leq x < 0.65 \\ 0.05 & 0.65 \leq x < 1.95 \\ 0.10 & 1.95 \leq x < 3.25 \\ 6.40 & 3.25 \leq x < 4.58 \\ 6.45 & 4.58 \leq x < 5.91 \\ 6.50 & 5.91 \leq x < 7.24 \\ 6.55 & 7.24 \leq x < 8.57 \\ 6.60 & 8.57 \leq x < 9.90 \\ 13.10 & 9.90 \leq x < 11.25 \\ 13.15 & 11.25 \leq x < 12.60 \\ 13.20 & 12.60 \leq x < 13.95 \\ 13.25 & 13.95 \leq x < 15.30 \\ 13.30 & 15.30 \leq x < 16.65 \\ 19.90 & 16.65 \leq x. \end{cases}$$

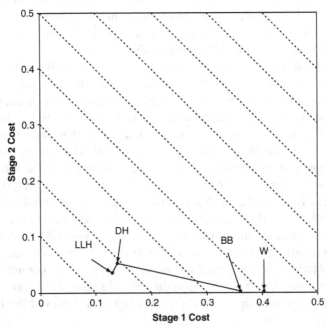

Fig. 8.26. Historical improvements on the Witsenhausen problem (benchmark: $k = 0.2$ and $\sigma = 5$) (Lee et al. 2001) © 2001 IEEE

The corresponding cost is $J_{LLH} = 0.167313205338$. This is the best-so-far solution in the past over-thirty years. In summary, we show the historical improvements on the Witsenhausen problem in Fig. 8.26, from which we can see J_{LLH} has a good balance between stage 1 cost and stage 2 cost. It was claimed that this is the "optimal" solution in the sense that all properties of the optimal solution have been discovered, any further improvement is relatively small and can only be achieved numerically, say by further dividing each step of function $f_{LLH}(x)$ into smaller steps and using local search to improve.

4.2 Application of OO for simple and good enough control laws

Although f_{LLH} is the best-so-far solution to the Witsenhausen problem, it is obviously more complex than the 1-step function f_W and f_{BB}, the 2-step function f_{DH}, and the following 3.5-step function $f_{3.5}$

$$f_{3.5}(x) = \begin{cases} 0 & 0 \le x < 3.25 \\ 6.5 & 3.25 \le x < 9.90 \\ 13.2 & 9.90 \le x < 16.65 \\ 19.9 & 16.65 \le x. \end{cases}$$

from which f_{LLH} was obtained by adding small segments to each step. Each time when a better $f(x)$ was reported, the incremental improvement becomes smaller and smaller (from $J_W = 0.4042$ to $J_{BB} = 0.3634$ to $J_{DH} = 0.1901$ to $J_{3.5} = 0.1714$ to $J_{LLH} = 0.1673$), and the function f becomes more and more complex. It seems that there is a trade-off between the performance of f and the complexity of f. Since we have not quantified the complexity of f, this statement is very informal. An interesting question is whether we can find simple f's with similar performance to the best-so-far solution. Though f_W, f_{BB}, f_{DH}, and $f_{3.5}$ are intuitively simpler than f_{LLH}, these functions were not obtained for being simple and good. It is not clear yet how to find a simple and good enough f in a systematic way. This is where the OO methodology of Chapter VI can help. In this subsection, we use the Kolmogorov complexity (KC) to measure the complexity of a function f. The KC of a function can hardly be calculated in general, but can be estimated through the OBDD-based representation of the function (where OBDD stands for Ordered Binary Decision Diagram). In Chapter VI, we combine OO and OBDD to get a systematic method of finding simple and good enough solutions. We will use this method to find a simple and good enough control law for the Witsenhausen problem. Comparing with the best-so-far function f_{LLH}, with minor performance degradation (within 5%), we reduce the complexity of f (i.e., the memory space to store f) by over 30 folds. Although in this specific example, most digital computers in engineering practice can store f_{LLH} with no difficulty, the importance of studying this problem is to demonstrate how to use the method introduced in Chapter VI to find simple and good enough designs, especially when the memory space is limited.

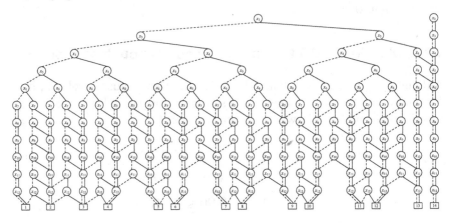

Fig. 8.27. One PROBDD that describes f_{LLH}. The 14 boxes in the bottom represent the 14 output in f_{LLH}, from 0.00 to 19.90, represented by 11-bit binary sequence

We start from quantifying the complexity of the best-so-far function f_{LLH}. As introduced in Chapter VI, KC supplies a measure of this complexity. The idea is to write a program to represent f_{LLH}. When the value of x is input, a computer should output the value of $f_{LLH}(x)$ by executing this program, which should work for all the values of x. The length of the shortest program that can represent f_{LLH} in this way is defined as the KC of f_{LLH}. $f(x)$ is defined over all the real numbers. However, digital computers have only finite input. Lee et al. showed that (Lee et al. 2001) it is reasonable to approximate $f(x)$ by only focusing on the domain of $[-5\sigma, 5\sigma]$, i.e., $-25 \leq x \leq 25$, because $x \sim N(0, \sigma^2)$ and the value of $f(x)$ for $x \notin [-5\sigma, 5\sigma]$ are very insignificant to the overall cost objective. In the following discussion, when representing function $f(x)$ by a program, we only consider input $0 \leq x \leq 25$. In f_{LLH}, the input has a resolution of 0.01, thus we need $\lceil \log_2(2500) \rceil = 12$ bits to encode the input. As for the output, we can similarly calculate that $\lceil \log_2(1990) \rceil = 11$ bits are needed to encode the output. As explained in Chapter VI, we can also represent f_{LLH} by a Partially Reduced OBDD (PROBDD) with 12-bit input and 11-bit output. We show one such PROBDD in Fig. 8.27. There are 182 nodes (excluding the 14 boxes in the bottom) in this PROBDD. Following the calculations introduced in Chapter VI, we know it takes 24192 bits[10] to store this PROBDD. The details of the calculation are shown in Table 8.13. (In the table, f_{sg} is a *simple* and *good enough* solution which will be discussed later.) The size 24192 is much less than 57500, the size of lookup table representation.

Exercise 8.3: In the look-up table representation, we describe f_{LLH} by listing all the $(x, f_{LLH}(x))$ pairs in sequence. Please tell why 57500 bits all together is needed.

Table 8.13. The complexity and performance of the milestone f's

f	Base (input)	# of input bits	Base (output)	# of output bits	# of nodes	# of boxes	# of rules (r)	d	4rd (bits)	J
f_W	–	0	1	3	0	1	6	4	96	0.4042
f_{BB}	–	0	0.0001	16	0	1	32	5	640	0.3634
f_{DH}	0.01	12	0.0001	17	32	2	132	7	3696	0.1901
f_{sg}	0.01	12	0.001	15	24	4	168	7	4704	0.1746
$f_{3.5}$	0.01	12	0.1	8	62	4	188	7	5264	0.1714
f_{LLH}	0.01	12	0.01	11	182	14	672	9	24192	0.1673

This justifies that PROBDD supplies a more compact representation of the function f_{LLH}. Since the KC of a function cannot be calculated in general, we suggest in Chapter VI to use "$4rd$" as an estimate of the KC of a function f, where r is the number of rules to implement the PROBDD that represents f, and d is the number of bits to encode each of the 4 elements in a rule. Following this measurement, we also estimate the KC of f_W, f_{BB}, f_{DH}, and $f_{3.5}$, and show the results in Table 8.13. We show the complexity and

---

[10] In (Jia 2006) and (Jia et al. 2006b), the complexity of f_{LLH} is estimated as 71311 bits, which is different from the results shown here. There are several reasons that cause the difference. First, (Jia et al. 2006b) used 15 bits to encode the input and 14 bits to encode the output. Second, due to more bits used in input and output, the PROBDD obtained in (Jia et al. 2006b) contains more nodes than the one shown in Fig. 8.27. Third, instead of using the formula "$4rd$" to estimate the complexity (Please refer to Chapter VI for more details), (Jia et al. 2006b) uses another formula to estimate the complexity. Although the values of the estimates are different, (Jia 2006) also estimates f_{sg}, which will be introduced later in this subsection, much simpler than f_{LLH}. At the end of this subsection, we will use an example to show how different computers may lead to different KCs. In practical application, this will not be a problem, as the computer is given and fixed.

performance of these functions in Fig. 8.28. This also justifies that $4rd$ is a reasonable estimate of the complexity of the function, because Fig. 8.28 shows that when the performance improves, the complexity also increases, which is consistent with our intuition.

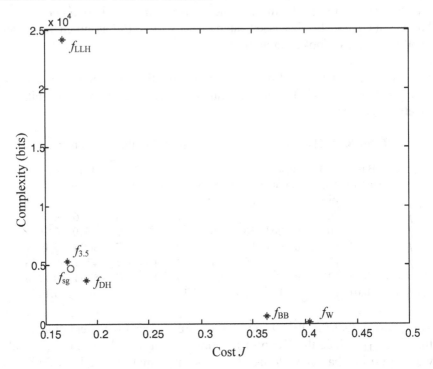

Fig. 8.28. The complexity and performance of the milestone f's

After comparing milestone solutions to the Witsenhausen problem, we will return to the main topic of this subsection: to find simple functions with good performances (say similar to that of f_{LLH}) using method introduced in Chapter VI. For this sake, we add a constraint $\hat{C}(f) \leq 10000$ where $\hat{C}(f)$ represents the estimate of the complexity of f as explained above. This constraint can be interpreted as the given memory space of 10000 bits. As explained above, we need 24192 bits to store f_{LLH} using PROBDD, and 57500 bits using lookup table, so f_{LLH} cannot be stored. Following the method introduced in Chapter VI, by randomly generating 1000 control functions satisfying that $\hat{C}(f) \leq 10000$, we randomly sample solutions to the Witsenhausen problem that can be stored within the

10000-bit memory space. Using numerical integration with a large step-size to estimate the cost J of each such function as in (Lee et al. 2001), we obtain an observed Ordered Performance Curve as shown in Fig. 8.29.

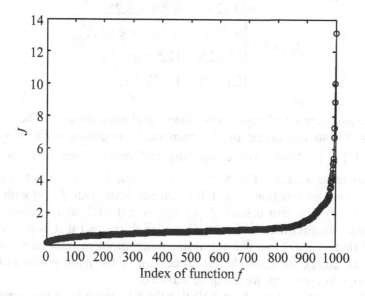

Fig. 8.29. The observed OPC of the functions that can be stored within 10000 bits in the Witsenhausen Problem (Jia et al. 2006b) © 2006 IEEE

This indicates that our problem belongs to the flat type of OPC. Also the noise level is estimated as 0.0056 after normalization. Suppose we want to find at least one of the top-5% designs that can be stored in 10000 bits. Using the UAP table in Section II.5, we calculate that we should select the observed top-37 designs. The OO theory ensures that there is at least one of the top-5% simple functions contained in the observed top-37 simple functions with a high probability. To test this, we run the above procedure by 100 replications. Each time, we randomly sample 1000 simple functions, and use long-time numerical integration to calculate the accurate performances. We find that there is always at least one truly top-5% simple function found each time, i.e., the observed alignment probability is 1 in the experiments.[11] There are actually 24.5 truly top-5% simple functions

---

[11] The top-5% here is w.r.t. all the simple functions. Since we do not sample complex functions (i.e., functions that need more than 10000 bits to store) in the experiments, we may not find a truly top-5% function (w.r.t. both simple and complex functions) in each replication.

found on average, with an average cost of 0.2096. The best function found in the 100 replications is

$$f_{sg}(x) = \begin{cases} 3.125, & 0 \le x < 6.25; \\ 9.375, & 6.25 \le x < 12.5; \\ 15.625, & 12.5 \le x < 18.75; \\ 21.875, & 18.75 \le x, \end{cases}$$

where the subscript "sg" represents "simple and good enough". The cost of $f_{sg}(x)$ is 0.1746 (estimated by the numerical integration with step size 0.001). $\hat{C}(f_{sg}) = 4704$. The complexity and performance of f_{sg} is also shown in Table 8.13 and Fig. 8.28. Although function $f_{sg}(x)$ is not as good as the best-so-far function $f_{LLH}(x)$, it is already better than $f_{DH}(x)$ with cost 0.1901, $f_{BB}(x)$ with cost 0.3634, $f_W(x)$ with cost 0.4042, and of course the best linear function with cost 0.96. Considering the fact that in the procedure of finding $f_{sg}(x)$, we did not utilize much problem information as in (Lee et al. 2001), for example, the placement of the jump points and the slight slope in each step, we are quite satisfied.

We have mentioned in Chapter VI that the KC depends on the computer that executes the program. Some computer has a longer list of commands. This allows us to use a shorter program to implement the same function. For example, in some computers the numbers 0.05, 0.10, 0.15, and 0.20 are not stored in the normal form, i.e., the binary numbers converted directly from the decimals. Instead, these computers record a base 0.05, and record the numbers as 1, 2, 3, and 4. In this way, we only need to save one decimal fraction accurately, thus save the memory space. This technique is commonly adopted in our laptops and desktops. By using this technique, we can further reduce the complexity of function f's from Table 8.13 to Table 8.14. In such a computer, f_{sg} only requires 600 bits. With minor performance degradation (within 5%), we save the memory space by over 30 folds (from 22176 bits to 600 bits). We also show the new complexity and performance in Fig. 8.30. In Fig. 8.30, f_{BB} dominates f_W in the sense that they have the same complexity but f_{BB} has better performance. f_{sg} dominates f_{DH} in both the complexity and the performance. The f_{LLH}, $f_{3.5}$, f_{sg}, and f_{BB} are the Pareto frontier.

Table 8.14. The complexity and performance of the milestone f's when allowing to change the base in input and output

f	Base (input)	# of input bits	Base (output)	# of output bits	# of nodes	# of boxes	# of rules (r)	d	4rd (bits)	J
f_W	–	0	5	1	0	1	2	3	24	0.4042
f_{BB}	–	0	3.9894	1	0	1	2	3	24	0.3634
f_{DH}	6.41	1	0.0001	17	1	2	70	6	1680	0.1901
f_{sg}	6.25	2	3.125	3	3	4	30	5	600	0.1746
$f_{3.5}$	0.05	9	0.1	8	48	4	160	7	4480	0.1714
f_{LLH}	0.01	12	0.05	9	182	14	616	9	22176	0.1673

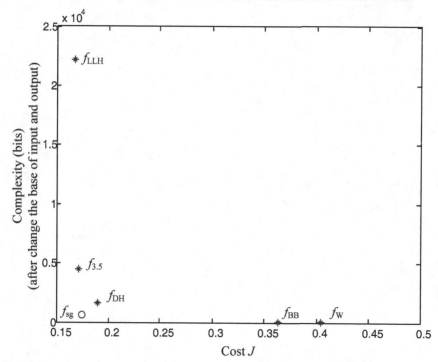

Fig. 8.30. The complexity and performance of the milestone f's after changing the base of input and output

4.3 Conclusion

In this section we have considered the famous Witsenhausen problem in team decision theory. By comparing the observed PDF of each subset, we find an easy way to discover several properties of good designs, using numerical experiments, thus can find the promising subsets efficiently. When applied in year 1999, this method finds a function for the Witsenhausen problem, which is better than the best function known by that time. Following the step-function formulation, a best-so-far function was found in 2001 by Lee et al. However, the solution seems involving a high degree of descriptive complexity. By combing OO and OBDD as introduced in Chapter VI, we are able to find simple and good enough functions with high probability. Applying this idea to the Witsenhausen problem, with minor performance degradation, we save the memory space by over 30 folds.

Appendix A Fundamentals of Simulation and Performance Evaluation

1 Introduction to simulation

There are many excellent textbooks on the subject of simulation (Yakowitz 1977; Fishman 1996; Gentle 2003; Landau and Binder 2000; Bartley et al. 1987). It is not the purpose of this chapter to repeat those materials. What are gathered here for convenience are simply the essentials we need for the purpose of this book, for quick review and reference, and for readers who have no previous exposure to the simulation literature. Readers familiar with the topic can just quickly note the headings and/or skim/skip this appendix.

Simulation is the electronic equivalent of a "pilot plant or laboratory mockup". It is a form of modeling. In modeling, of course, there is always the question of how faithful and how much detail we wish to incorporate in the model. We shall not be concerned with this issue here. We shall assume that whatever simulation model we are finally using in this book is, it is as faithful an electronic copy of the real system as possible. We are only concerned with two aspects of simulation modeling.

1. the laboratory aspect - the software which includes general purpose algorithms and interfaces, e.g., the Generalized Semi-Markov Processes (GSMP) model (see Appendix B for an explanation) and the GUI object oriented features.
2. the statistical aspect - the analysis of output data as a statistical experiment. This is performance evaluation, and if feasible, performance optimization. This is the goal of this book.

Mathematically, the performance of a system (as discussed in Chapter I) is measured by $J(\theta,\xi) = E[L(x(t; \theta,\xi)]$, where θ is the vector of system parameters, ξ all the randomness in the system, $x(t; \theta,\xi)$ a sample path of the system, L a functional measuring the performance of the system under the sample path $x(t; \theta, \xi)$. The system can be mathematically and conceptually captured by a Generalized Semi-Markov Process (GSMP) as described in Appendix B. A simulation model of a system is then simply a software implementation of the GSMP model. A particular implementation is the so called event scheduling approach to discrete event simulation.

The event scheduling approach to simulation modeling can be explained via a diagram (Fig. A.1).

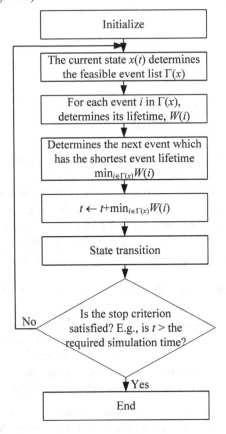

Fig. A.1. Flow chart of event scheduling based simulation

Note time steps forward from event to event in this approach in contradistinction to the integration of differential equations in **Continuous Variable Dynamic Systems (CVDS)** where time marches on in small increments of Δt.

To facilitate the implementation of Fig. A.1, we need various software ingredients:

        registers for state, time, scheduled (future) event list,
        routines for initialization, state transition, update time,
        statistics gathering, output report, and random variable generation,

main program which models the DEDS (user written),
modern features, e.g., animation, object-oriented programming, etc.

Many excellent simulation languages and software exist to do these (e.g., http://www.imaginethatinc.com). We shall not go into them here.

2 Random numbers and variables generation

Samples of random variables distributed according to specifications are continuously needed in simulation implementation, i.e., the ξ in $E[L(x(t; \theta, \xi))]$. As we will see in the following discussion, generating independent samples from a uniform distribution over the unit interval $(0,1)$ is the foundation to obtain samples for more complicated distributions. Most commonly used random number generator is introduced in Subsection 2.1.

2.1 The linear congruential method

$$x_{n+1} = \mathrm{mod}_M \left[a x_n + b \right] \text{ and } u_{n+1} = x_{n+1}/M . \tag{A.1}$$

When a, b, M takes appropriate values, u_1, u_2, ..., u_n,... form a uniform distribution. Although Eq. (A.1) looks quite simple, parameters in this generator must be carefully chosen in order to produce usable samples.

Example 1 Let $a = 2$, $b = 1$, and $M = 16$. Using Eq. (A.1) and various x_0's, we get

x_0	1	2	4	6	8	10	12	13	14
x_1	3	5	9	13	1	5	9	11	13
x_2	7	11	3	11	3	11	3	7	11
x_3	15	7	7	7	7	7	7	15	7
x_4	15	15	15	15	15	15	15	15	15
x_5	15	15	15	15	15	*All sequences get stuck*			
x_6	•	•	•	•	•	*after the initial transients!*			

Example 2 Let $a = 3$, $b = 0$, and $M = 16$. Similarly, we have

$$x_0 = \begin{array}{ccccc} 1 & 2 & 4 & 5 & 8 & 10 \\ 3 & 6 & 12 & 15 & & 14 \\ 9 & & & 13 & & \\ 11 & & & 7 & & \end{array}$$

Note that, depending on the initial seeds, the sequences get into cycles with different periods. However, none of the sequences produces the maximal period of 0-15 (in other words, none of the sequences produces all the integers 0, 1,...14, and 15.).

Example 3 Let $a = 1$, $b = 3$, and $M = 16$. Starting with *any* seed, we get the maximal period and the sequence [. . . , 1, 4, 7, 10, 13, 0, 3, 6, 9, 12, 15, 2, 5, 8, 11, 14, 1, ...] this time. This is nice. However, a plot of the sequence vs. time shows high correlation among successive numbers in the sequence as illustrated in Fig. A.2. Thus the numbers in the sequence are not at all independent.

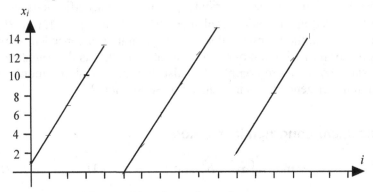

Fig. A.2. Plot of pseudo random sequence

Example 4 Let $a = 5$, $b = 3$, and $M = 16$. Once again we get a sequence of maximal period with any seed, [. . . . , 1, 8, 11, 10, 5, 12, 15, 14, 9, 0, 3, 2, 13, 4, 7, 6, 1, . . .]. A similar plot as in Fig. A.3 shows a reasonably random looking sequence.

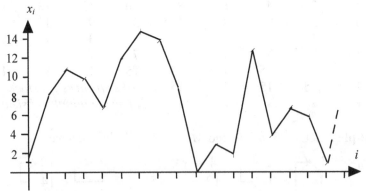

Fig. A.3. Yet another plot of pseudo random sequence

Thus, **periodicity and correlation** are important issues in random number generation. Good random generators should have long period and least correlation. More generally, the quality of random number generators should be quantified and tested for most serious work. This includes not only the mathematical analysis of properties of the generators, but also a set of empirical statistical tests on the samples produced by the generators. These tests try to detect empirical statistical properties of a sequence against the null hypothesis H_0 - "the samples are realizations of i.i.d. $U(0,1)$ random variables." For the last word see (Tezuka 1995) or (Gentle 2003).

There are a number of methods used for random variable generation with general distribution. We focus on two methods: inverse transformation and rejection.

2.2 The method of inverse transform

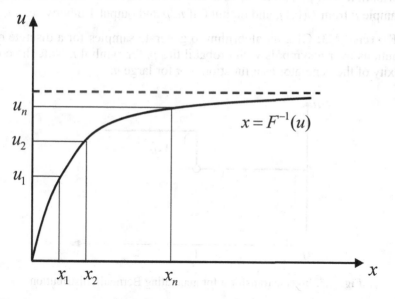

Fig. A.4. Inverse transform for generating $F(x)$-distributed random variables

As Fig. A.4 shows, in this method, a sequence of samples $u_1, u_2, \ldots$ from uniform distribution $U(0,1)$ are first generated, the sequence of numbers $F^{-1}(u_1), F^{-1}(u_2), \ldots$ are used as the samples $x_1, x_2, \ldots$ for the random variable whose cumulated distribution function is given by $F(x)$, and $F^{-1}(u)$ is the inverse function of $F(x)$. To see why this method works, consider the probability

$$\text{Prob}[x \le a] = \text{Prob}[F^{-1}(u) \le a] = \text{Prob}[u \le F(a)] = F(a),$$

where the last equality is by virtue of the uniform distribution of the random variable u.

Example To generate a random number contains exponential distribution, let $u = F(x) = 1 - \exp(-\lambda x) \Rightarrow \ln(u-1) = \lambda x$ or $x = (1/\lambda)\ln(u-1)$, where u is a random number that contains uniform distribution $U(0,1)$.

Exercise A.1: How can we generate samples for Weibull distribution $F(x)=1-\exp(1-(x/\beta)^\alpha)$, where $\alpha>0$ and $\beta>0$ are two parameters?

Exercise A.2: What are the possible limitations of this method? Can it be extended to multi-dimensional distributions?

The inverse function method also applies to discrete distributions. It is essentially a table lookup procedure. An example of a common and very simple application of this technique is to generate samples for Bernoulli distribution with parameter p, as is shown in Fig. A.5. That is, to generate a sample u from $U(0,1)$, and output 0 if $u \leq p$ and output 1 otherwise.

Exercise A.3: Give an algorithm to generate samples for a discrete distributions over n symbols with probabilities p_i for symbol i. Note the complexity of the generator as a function of n for large n.

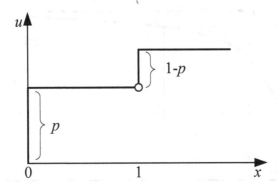

Fig. A.5. Inverse transform for generating Bernoulli distribution

2.3 The method of rejection

If we know the probability density function $f(x)$ of a random variable x with finite support (being non-zero only in some finite interval $[a,b]$), we can use the following method, called rejection method, to generate samples of the random variable. The basic idea is that, by selectively discarding samples from one distribution, we can make the remaining samples stochastically equivalent to samples from a different distribution. Suppose we know that c is an upper bound of $f(x)$.

Step (1) generate u, a sample of a uniform random number U on $[a,b)$;
Step (2) generate v, a sample of a uniform random number V on $[0,c)$;
Step (3) if $v \le f(u)$, accept u as a sample of X; otherwise go to step (1).

Pictorially, steps (1) and (2) generate samples uniformly on the rectangular area defined by sides ac and ab. Step (3) simply throws away the samples in the shaded region as is shown in Fig. A.6. Those in the unshaded region are retained to yield samples of x with density function $f(x)$.

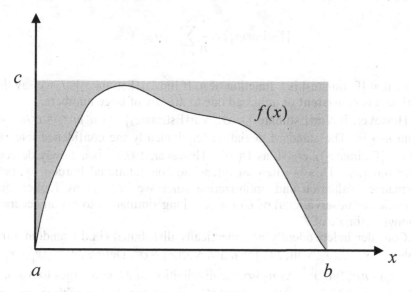

Fig. A.6. Rejection method of random variable generation

Exercise A.4: What are the possible disadvantages of this method? Can it be extended to multi-dimensional distributions?

A set of useful facts associated with exponential distribution, discrete event systems and particularly queuing systems are as follows:

The **Memoryless** triangle: the following three statements are equivalent and each one implies the other and vice versa –

1. The inter-event times are exponentially distributed
2. Event arrivals are Poisson distributed
3. The process is Markov and memoryless

Exercise A.5: Please verify this.

3 Sampling, the central limit theorem, and confidence intervals

In analyzing outputs of a simulation experiment, we often need to estimate the mean of a random variable, L, i.e., $J=E[L(\theta,\xi)]$. Consider

$$[\text{Estimate}_1] = \text{any sample of } L.$$

If $E[\text{Estimate}_1]=J$, we say such an estimate is **unbiased.** Consider again

$$[\text{Estimate}_2] = \frac{1}{n}\sum_{i=1}^{n} L(\theta,\xi_i).$$

Note that [Estimate$_2$] is a function of n. If $\lim_{n\to\infty}[\text{Estimate}_2]=J$, we say the estimate is **consistent** or unbiased due to the law of large numbers.

However, if $\text{Var}[\text{Estimate}_1] = \sigma^2$, $\text{Var}[\text{Estimate}_2]= (1/n^2)n\sigma^2 = \sigma^2/n \to 0$ with $n\to \infty$. The standard deviation (equivalently the confidence interval of the [Estimate$_2$]) $\sigma \to 0$ as $1/(n)^{1/2}$. However, $1/(n)^{1/2}$ is a slowly decreasing function. This is often an infeasible computational burden for performance evaluation and optimization since we need many replications (samples or observations) of L or a very long simulation to get an accurate enough estimate of J.

Consider independently and identically distributed (i.i.d.) random variables $x_1, x_2, \ldots, x_n$, with $E[x_i] = \mu$ and $\text{Var}[x_i] = \sigma^2$. Define $M_n = [(x_1 + x_2 + \ldots + x_n) - n\mu]/(n\sigma^2)^{1/2}$. As $n\to\infty$, the distribution of M_n converges to *Normal distribution* $N(0,1)$, i.e., the normal (Gaussian) distribution with mean zero and unit variance. This is known as the **Central Limit Theorem (CLT)**. The significance of CLT for experimental work lies in the fact that it enables us to predict the error of sampling. For example, suppose we take n samples of a random variable with mean μ. We may use $\bar{x} \equiv (x_1 + x_2 + \ldots + x_n)/n$ as an estimate for the unknown mean μ. Then M_n is the normalized error of the estimate. For large n, we can use standard tables for Gaussian random variables to calculate $P\equiv\text{Prob}[-t<M_n<t]$, which is the probability that the error of the estimate for μ lies in $[-t, t]$. For example, if $t =1.96$, we get $P=0.95$; i.e., we are 95% confident that the interval $[\bar{x} -t\ (\sigma^2/n)^{1/2}, \bar{x} + t\ (\sigma^2/n)^{1/2}]$ contains the unknown mean μ. In turn, for a specified confidence and interval size, we can calculate how many trials of the experiment are needed.

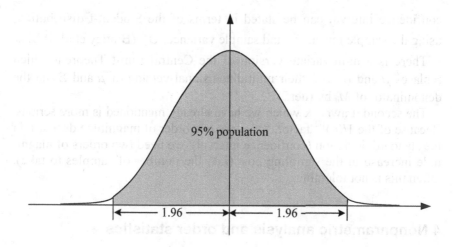

Fig. A.7. The density of standard normal distribution $N(0,1)$

With probability 0.95 the interval $\left[\bar{x}-1.96\sqrt{\dfrac{\sigma^2}{n}}, \bar{x}+1.96\sqrt{\dfrac{\sigma^2}{n}}\right]$ contains the unknown mean μ, as is shown in the Fig. A.7. We have $\sum\limits_{i=1}^{n} x_i / n - \mu \sim N\left(0, \sigma^2 / n\right)$ or

$$y \equiv \frac{\left(\sum\limits_{i=1}^{n} x_i - n\mu\right)}{\sqrt{\sigma^2 / n}} \sim N(0,1).$$

The above confidence interval formula suffers from two drawbacks. **First**, it requires the knowledge of the variance of the random variable, σ^2. It hardly seems reasonable that we can know the value of σ when not even the mean μ is known. The common practice is to replace σ^2 with the sample variance,

$$\sigma_s^2 = \frac{1}{n}\sum_{i=1}^{n}(x_i - \bar{x})^2 \text{ or } \sigma_s^2 = \frac{1}{n-1}\sum_{i=1}^{n}(x_i - \bar{x})^2,$$

in which case the formula is only approximate. However, if we do know that the random variable in question is Gaussian, an exact formula for the confidence interval can be stated in terms of the Student-t distribution, using the sample mean, $\bar{x}$, and sample variance, σ_s^2 (Bratley et al. 1987).

There is a multivariate version of the Central Limit Theorem which replaces μ and σ with their multidimensional version of μ and Σ and the denominator of M_n by $(\det\Sigma)^{n/2}$.

The second drawback which we have already mentioned is more serious. Because of the $1/(n)^{1/2}$ factor, for every one order of magnitude decrease in the standard deviation (confidence interval), we need two orders of magnitude increase in the sampling cost (i.e., the number of samples to take). Often this is not tolerable.

4 Nonparametric analysis and order statistics

Suppose you take n i.i.d. samples of an arbitrary random variable. Now you order the samples by magnitude into $x_{[1]} < x_{[2]} < x_{[3]} < \ldots < x_{[n]}$. The theory of order statistic (David et al. 2003) says that these order statistics on average divide the population into $n+1$ equal parts. Furthermore, we can calculate the probability for what % of the population is contained below, above or between any one or two order statistics. This is something one gets for free in any statistical experiment, including simulation.

5 Additional problems of simulating DEDS

A large part of output analysis of simulation consists of reducing the variance of estimate and shorten the computational burden without sacrificing unbiasedness or consistency.

(i) Transients

Performance indices of interests are usually long term measurements for systems operating in stationary states. However, the stationary states are not known before the simulation. Short simulation starting from a given initial state will lead to biased estimation of the desired performance indices, no matter how many replications are executed.

(ii) Correlation

If we are estimating parameters of stationary distributions, the data we collected are usually correlated instead of forming i.i.d. random variables. This may not make the estimator unbiased, and can increase the variance of the estimation significantly.

(iii) Antithetic random variables

This is a variance reduction technique by introducing negatively corre-lated random variables in constructing estimators. Its basic idea comes from the following observation. Suppose A and B are two random vari-ables having the same expectation θ. Then $E[(A+B)/2] = \theta$. So, $(A+B)/2$ is an estimator for θ. If A and B are negatively correlated, that is $\text{Cov}(A,B)<0$, the estimator $(A+B)/2$ is better than when A and B are independent. This can be seen from the fact that $\text{Var}[(A+B)/2] = (\text{Var}[A] + \text{Var}[B] + 2\text{Cov}(A,B))/4$. The common practice is to use complementary samples to balance the random effect. For example, if we sample from a uniform dis-tribution in $(0,1)$, both X_k and $1-X_k$ should be used.

(iv) Regenerative cycles

On a sample path of a stochastic process, if there are existence of times, usually random, from which onward the future of the process is a probabil-istic replica (or copy) of the original process, we call the durations between these times regenerative cycles. From the probabilistic viewpoint, all cycles are statistically equivalent and constitute a genuine sample of the simulation experiment. Thus, the parameters of the stationary distribution for processes having regenerative cycles can be estimated in a single sam-ple path, in contrast to multiple replications for general situations in which no such regenerative cycles exists. We avoid the problem of "setup" and "initialization" for each of the replications.

(v) Common random variables

This variance reduction technique is useful when we estimate the differ-ence between two random variables (recall that the sampled values for per-formance indices are random variables). For example, we are interested in estimate $E[A-B]$ for A and B by Monte Carlo simulation. We have $\text{Var}[A-B]=\text{Var}[A]+\text{Var}[B]-2\text{Cov}(A,B)$. If we generate samples for A and B in a way such that they are positively correlated, for example, $A=f_1(X)$ and $B=f_2(X)$ where both f_1 and f_2 are increasing functions of a common random variable X, $\text{Var}[A-B]$ will be less than when A and B are independent. The common practice is to use the same random numbers under different con-figurations when we compare their effects.

(vi) The use of warm-up period

The main difficulty to estimate steady-state parameters is that we need to obtain long sample paths. One reasonable way to speed up the conver-gence of estimation on the steady-state parameters is to omit the initial part of sample paths which depend more on the initial states. This is known as *initial data deletion* or as *warming up* the simulation.

(v) Separate batches

In estimating steady-state parameters, one way is to take non-over-lapping batch means (NOBM), which is to divide simulation outputs of a single replication, after deletion of a warm-up period, into k adjacent non-overlapping batches, each of size m. If the batch size m is sufficiently large so that the batch means are approximately i.i.d. normal random variables, we can apply CLT to estimate steady-state parameters. The advantage of this method is to use a single replication, so that only one warm-up period is deleted.

6 The alias method of choosing event types

In Fig. A.8, we see that an important step in discrete event simulation is the generation of various events. This is equivalent, from a simulation viewpoint, to the standard problem of the generation of discrete random variables according to arbitrary distributions. This can become time-consuming when the domain of the random variable is large (see Exercise A.3). One **efficient** way of obtaining a random variable distributed over the integers $1,2,...,n$ with probabilities p_i, $i=1,2,...,n$, is the alias method (Walker 1974). This method can be used to further reduce the computation effort in simulation, e.g., in the standard clock simulation approach (more details in Section VII.2) in determining the event type at every transition instant (See (Cassandras and Lafortune 1999)). The method requires only one uniformly distributed variable, one comparison, and at most two memory references or table lookups per sample. It is thus **independent** of the size of the possible event list, an important advantage in the simulation of large systems via the standard clock approach. However, this method requires pre-computing two tables of length n, which is a one-time effort.

Our summary of the alias method is based on (L'Ecuyer 2004). To explain the idea of alias method, we consider a bar diagram of the distribution, where each index i has a bar of height p_i=Prob[X=i]. The idea is to "equalize" the bars so that they all have height $1/n$, by cutting-off bar pieces and transferring them to other bars. This is done in a way that in the new diagram, each bar i contains one piece of size q_i from the original bar i and one piece of size $1/n-q_i$ from another bar whose index j, denoted as $A(i)$, is called the alias value of i. The setup procedure initializes two tables, A and R, where $A(i)$ is the alias value of i and $R(i)=(i-1)/n+q_i$. To generate X, we generate $U\sim U[0,1]$, denote $I=\lceil nU \rceil$, where $\lceil \bullet \rceil$ is the ceiling function, and return X=I if U<$R(I)$ and X=$A(I)$ otherwise.

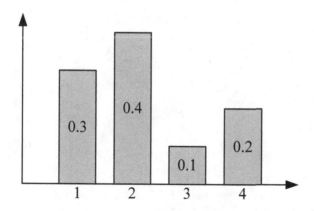

Fig. A.8. A random variable with discrete distribution

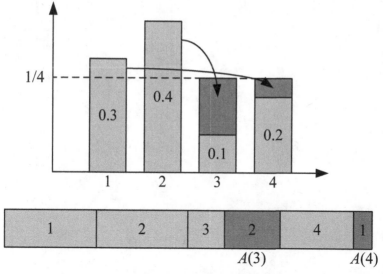

Fig. A.9. The alias method to generate the random variable

Consider the random variable whose probability distribution is given by Fig. A.8, namely it takes 4 possible values {1,2,3,4} with probability p_1=0.3, p_2=0.4, p_3=0.1, p_4=0.2. Choose q_1=1/4, q_2=1/4, q_3=0.1, q_4=0.2. The tables A and R can be constructed as $A(1)$=1, $A(2)$=2, $A(3)$=2, $A(4)$=1; $R(i)$=$(i$-1)/n+q_i, $R(1)$=(1-1)/4+q_1=1/4, $R(2)$=(2-1)/4+q_2=1/2, $R(3)$=(3-1)/4+q_3=0.6, $R(4)$=(4-1)/4+q_4=0.95. This selection is shown in Fig. A.9. For example, to generate X, we generate U~U[0,1]. Suppose U=0.7. Then I=$\lceil nU \rceil$=$\lceil 4 \times 0.7 \rceil$=3. Because 0.7>$R(3)$, return X=$A(3)$=2.

Appendix B Introduction to Stochastic Processes and Generalized Semi-Markov Processes as Models for Discrete Event Dynamic Systems and Simulations

1 Elements of stochastic sequences and processes

A Stochastic sequence is simply an indexed collection of random variables $x_1, x_2, \ldots x_i, \ldots$ specified by the complete joint density function $p(x_1, x_2, \ldots x_i, \ldots) \equiv p(x)$ among all the random variables $x = [\, x_1, x_2, \ldots x_i, \ldots]^1$. (In more general cases, each component of x, x_i, could be a vector itself (say m dimensional). If there are n components in x, the stochastic sequence x looks like an m-by-n matrix.) However, $p(x)$ is an n-dimensional function representing a tremendous amount of information and computationally infeasible to deal with (Ho 2005). Thus to mitigate the burden and provide more structures, we introduce below a series of definitions which specialize the joint density function $p(x)$.

First, we can provide an approximate characterization of $p(x)$ by specifying its first two moments:

$$\bar{x}_t = E[x_t]$$

and

$$R(t,\tau) = E\left[\left(x_t - \bar{x}_t\right)\left(x_\tau - \bar{x}_\tau\right)\right] = \int \left(x_t - \bar{x}_t\right)\left(x_\tau - \bar{x}_\tau\right) p(x) dx.$$

In the case of $t = \tau$, $R(t,\tau)$ becomes the covariance matrix of x_t, Σ_t. If x_t is a single random variable instead of a vector, Σ_t is simply the variance of x_t, $\mathrm{var}[x_t]$. If the density function $p(x_1, x_2, \ldots x_i, \ldots) \equiv p(x)$ is the same as $p(x_{t+1}, x_{t+2}, \ldots x_{t+i}, \ldots)$ for all t, we have a **stationary** sequence. If we only require

---

[1] This definition is good for finite sequence. For a collection of infinite many of random variables, a stochastic process should be specified by complete joint distributions among random variables over all possible finite set of indices.

$\bar{x}_t = \bar{x}_{t+\tau}$ and $R(t,\tau) = R'(t-\tau)$ for all t and τ, the sequence is said to be **wide sense stationary**.

Another approach to simplify and characterize a stochastic sequence is to specialize to a **purely random sequence** which requires

$$p(x) = \Pi_i \, p(x_i)$$

as the product of one-dimensional functions. The sequence then consists of independent random variables. If furthermore the density functions of all x_t are identical, we then have what is known as an **i.i.d. (independently and identically distributed)** sequence. In engineering terms, it is called a **white noise** sequence[2] when the mean is zero.

A **Gaussian sequence** is one summarized by

$$p(x) \sim N(\bar{x}, \Sigma),$$

where each component of x, x_t may be a vector itself with $p(x_t)$ $\sim N(\bar{x}_t, \Sigma_t)$.

Exercise B.1: Please explain every component of $\bar{x}$ and Σ when each component of x, x_t, is a vector.

A **Markov sequence** is when the joint density function has the additional property $p(x_{t+1}/x_t) = p(x_{t+1}/x_t, x_{t-1}, x_{t-2}, \ldots)$, which implies that we only need the initial density function, $p(x_0)$, and the transition density function, $p(x_{t+1}/x_t)$, to specify completely the stochastic sequence.

Exercise B.2: Translate the above sentence into mathematical terms.

Exercise B.3: Prove that for a Markov sequence $p(x_{t+1}/x_t, x_{t+2}) = p(x_{t+1}/x_t, x_{t-1}, x_{t-2}, \ldots, x_{t+2}, x_{t+3}, \ldots)$.

Exercise B.4: How do you characterize a Stationary Gauss-Markov sequence using the definitions above?

Exercise B.5: (For readers familiar with control theory) How do you represent a general Gauss-Markov sequence using difference equations and white noise sequences?

If, in addition, the state of the Markov sequence is discrete and finite, we have a Markov Chain and the Kolmogorov–Chapman equation

---

[2] More precisely, it is known as strong white noise. Some references refer to white noise as the wide sense stationary sequences with zero autocorrelation $R(t, \tau)$.

$$p(x_{t+1}) = \int p(x_{t+1}/p(x_t)) p(x_t) dx_t$$

becomes

$$\pi_{t+1} = \pi_t P$$

where

$$P_{ij} \equiv \text{Prob}[x_{t+1} = j \, / \, x_t = i]$$

and

$$\pi_t \equiv [\text{Prob}[x_t = 1], \text{Prob}[x_t = 2], \dots, \text{Prob}[x_t = N]],$$

where $\{1, 2, \dots, N\}$ is the state space.

For the above discussion, we implicitly assume that time index marches on uniformly in discrete time, e.g., $-2, -1, 0, 1, 2, \dots$. If, on the other hand, the indexing variable t is continuous, the number of index variable becomes infinite and uncountable. Conceptually we pass into the realm of **stochastic processes**. It involves mathematical machinery beyond the scope of this book to treat stochastic process rigorously. However, for all practical and computational purposes, we can restrict ourselves to stochastic sequences involving only a finite number of indexed variables. Moreover, the indexing variable t may not be uniformly spaced. In fact, "t" often represents "event times" when something happens, e.g., x_t jumps from one value to another at time t. In this case, we need to specify the times when the indexing take place. In other words, **the sequence of random variables carries two dimensions, one for the usual x, the other specifying the durations of time increment or time to the next event, and we enter the realm of the class of stochastic processes (sequences) that are special to Discrete Event Dynamic Systems (DEDS).**

If we concentrate on the density of the distribution of the time intervals between state transitions, $p(\tau)$ and simply treat the instants of transition (indexing) as an "event", we have a **renewal process** if the time intervals between events are independently identically but otherwise arbitrarily distributed. If furthermore, the common distribution is exponential, i.e.,

$$p(\tau = t) \sim \exp(-\lambda t),$$

we have a Poisson process, sometimes referred to as a **birth** or **death** process with rate λ which is a special type of **Markov** process[3]. On the other hand, in **Markov chains** with state transitions, the parameter λ_{ij} of

---

[3] This notation here is somewhat unfortunate but entrenched. Here the word "Markov" is applied to the distribution of time interval between events. If on the other hand, we also have Markov state transition at each event instant (when considered as a birth or death process), we really have a **doubly Markov process**.

the exponential distribution $\exp(-\lambda_{ij}t)$ may depends on the pair of states (i,j) where the state transition happens.

A discrete state continuous time process having Markovian state transition and renewal state transition interval is called a **Semi-Markov process (SMP)**[4], i.e., the state transitions x are Markov (**imbedded Markov Chain**) but the time intervals between events are not (generally rather than exponentially distributed)[5].

Semi-Markov processes are a special class of more general processes known as Generalized Semi-Markov process (GSMP). The following summary of GSMP is due to W. Whitt (Whitt 1980). A GSMP moves from state to state with the destination and duration of each transition depending on which of the several possible events associated with the occupied state occurs first. Several different events compete for causing the next jump and impose their own particular jump distribution to determine the next state. An ordinary Semi-Markov process (SMP) is the special case in which there is only *one* event associated with each state. At each transition of a GSMP, new events may be scheduled. For each of these new events, a clock indicating the time until the event is scheduled to occur is set by an independent chance mechanism. An event which is scheduled but does not initiate a transition is either abandoned or it is associated with the next state, and its clock just continues running.

If the events of the imbedded Markov chain in a Markov process are simply **births and deaths** with rates λ and μ, i.e., the intervals between events are exponentially distributed, then we have the process characterized by

$$p_{ij} = 0 \text{ for } |j-i| > 1.$$

If in addition, $p_{ij} = 0$ for $j < i$, we have a pure birth process, also known as a **Poisson** process. More generally, we can have state transition taking place continuously rather than discretely (e.g., via a stochastic differential equation), then we are in the realm of stochastic system theory which requires separate mathematical treatment not covered in this book. The following diagram (Fig. B.1) from the well-known book by Kleinrock (p. 25, vol. 1, (Kleinrock 1975)) visually captures the relationship between various stochastic sequences/processes. In this diagram, for example, the random walk $p_{ij} = q(j-i)$ means that the state transition probability p_{ij} only depends on the difference between the indexes of the two states.

---

[4] We always assume that the SMP generated by the Markov renewal process is time-homogenous as in (Çinlar 1975).

[5] Note, like in Markov Processes, the distributions $p(\tau)$ of the time intervals between state transitions in Semi-Markov processes are general and dependent on the pair of states where the state transition happens.

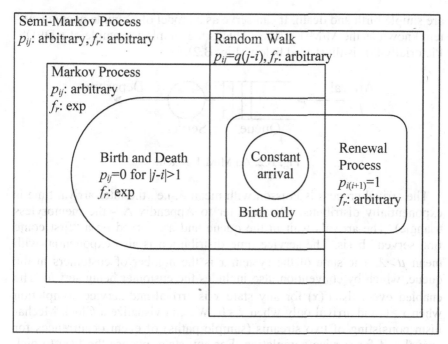

Fig. B.1. Relationship between various stochastic processes (Kleinrock 1975)
© 1975 John Wiley and Sons

Lastly, all the stochastic processes we have mentioned can be viewed as a special case of Markov sequences with *enlarged* state space. So long as the dependence on the past is finite, we can define the entire finite past including the present as the new enlarged state vector. The sequence of such vectors will be Markov. Other possibilities also exist. For example, if we regard the combination of process state and clock readings of all competing events (such as the arrival and departure in a queuing system) as general state x_n and interpret n as the counter of the occurrence of events. For such general states, we still have $p(x_{n+1}/x_n) = p(x_{n+1}/x_n, x_{n-1}, \ldots, x_0)$. The transition probability $P(s,A) = \text{Prob}[x_{n+1} \in A / x_n = x]$ is known as **transition kernel** for the corresponding process.

2 Modeling of discrete event simulation using stochastic sequences

Stochastic sequences discussed and described above can be used as models for the various Discrete Event Dynamic System (DEDS) and their simulations. We start from a doubly Markov process where the state transitions are simple birth and death. It can serve as a model of a simple queuing system known as the M/M/1 queue[6]. This is the simplest example of a DEDS. Pictorially this is illustrated below (Fig. B.2).

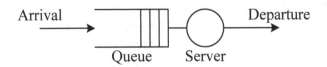

Fig. B.2. An M/M/1 system

The arrival process is Poisson with mean λ, i.e., the inter-arrival time is exponentially distributed (Please refer to Appendix A - the memoryless triangle). The arrivals wait in the queue and are served on a "first come first served" basis. The service time distribution is also exponential with mean $\mu > \lambda$. The state of the system x is the number of customers in the queue, which by convention also includes the customer being served. The enabled event list $\Gamma(x)$ for any state x is arrival and service completion when $x > 0$ and arrival only when $x = 0$. We can visualize a Clock Mechanism consisting of two streams (sample paths) of event occurrences for arrival and for service completion. For any state, we use the $\Gamma(x)$ to pick out and schedule the next event. A block diagram below (Fig. B.3) helps us to visualize the process graphically. (See also Fig. A.1 in Appendix A.)

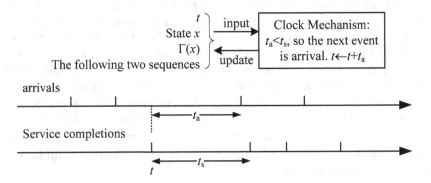

Fig. B.3. The clock mechanism for the M/M/1 system in Fig. B.2

We can generalize this process which leads to the Generalized Semi-Markov Process (GSMP) model of Discrete Event Dynamic System (DEDS),

---

[6] The "M" notation here stands for Markov arrival and service times of the simple queuing system.

which involves many streams of event occurrences and more general state transition function,

$$x_{next} = f(x_{now}, \text{triggering event, randomness}).$$

This is a good model for the behavior of many complex human-made systems and a simulation experiment of such systems.

We first introduce the **Untimed version** of a DEDS model, which contains the following ingredients.

State: $x \in X$ — a finite set = all possible states of the world
Event: $e \in \Gamma$ — a finite set = all possible happenings
Enabled Events: $\Gamma(x)$ — a subset of Γ depending on the state x
State Transition: $x_{new} = f(x_{now}, e_{now}, \text{randomness})$.

Alternative representation of the state transition function is

1. A Markov Chain Matrix $P^{(e)}$ with elements $P_{ij}^{(e)}$ = probability of transition from state i to state j when event e occurs. $P^{(e)}$ is a $|X| \times |X|$ matrix. We have one such matrix for each event.
2. A state table — more often used when the transition is deterministic, i.e., tabular data of $x_{new} = f(x_{now}, e_{now})$

A major drawback of this general model (often called automaton, or finite state machine, or Markov chain model) is that all structural information is not retained, e.g., in a simple queuing system, the state can only increment and decrement by one, so the state transition probability matrix can only be tri-diagonal.

Another problem which is common with discrete states is the combinatorial explosion of the size of the state space. This size often makes computation infeasible

Example Consider a serial queuing network which has N servers, each with M limited queue spaces for waiting tasks. The servers can be working, broken down, blocked if downstream queue-space is full, or starved if the upstream buffer is empty.

Exercise B.6: How many states are there for such a DEDS? What practical application this queuing network can represent?

Now we add "time" to the above model to create **a Timed version**. For readers versed in control theory, the timed model can be taken as the analog of $dx/dt = f(x,u,t)$ for DEDS (Bryson and Ho 1969).

For each enabled events in state x, we endow it with a lifetime or clock reading, $c(e_i)$, $e_i \in \Gamma(x)$. The clock readings run down at unit rate. An event is said to occur when the clock reading reaches zero. Occurrence of the event may trigger a state transition. The new state may enable additional events, each with new life times. The remaining life time of the other events also becomes the new life times in the next state if they are not disabled or interrupted. All these events again compete for termination, and the process repeats.

The informal description above can be visualized in Fig. B.4 below and formalized by using a Generalized Semi-Markov Processes framework[7]. The process is "Markov" because the state transition is a Markov Chain; it is "semi" because the inter-event times are arbitrarily distributed (with exponential inter-event time distribution, we'll simply have a Markov process); and finally, it is "generalized" because we can have several event processes concurrently going on to model a whole system.

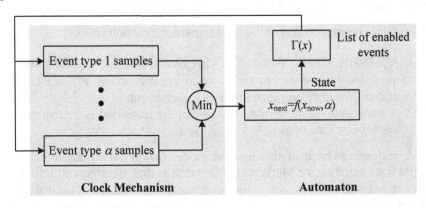

Fig. B.4. The timing and structural parts of the GSMP model

We define (Until the end of this section a great deal of notations are used to explain and formalize what basically is a simple notion as illustrated in Fig. B.4 above and on page 294 or 595 of (Cassandras and Lafortune 1999))

τ_n = the epoch of the nth state transition

a_n = the nth event

x_n = the nth state visited by the process: $x_n = x(\tau_n^+)$, where the superscript "+" represents the moment right after the nth state transition happens

c_n = the vector of clock readings at τ_n^+

---

[7] For readers who are not interested in mastering a great deal of notations, the mathematical text below can be skipped on a first reading.

$c_n(\alpha)$ = at τ_n, the time remaining until α occurs, provided $\alpha \in \Gamma(x_n)$, the feasible event list associated with x_n

$N(\alpha,n)$ = the number of instances of α among $a_1, \ldots, a_n$[8]

Also let $\omega_1 = \{Y(\alpha,k), \alpha \in \Gamma, k=1,2,\ldots\}$ and $\omega_2 = \{U(\alpha,k), \alpha \in \Gamma, k=1,2,..\}$ be two doubly indexed sequences of independent random variables. For each k, $Y(\alpha,k)$ is distributed[9] according to ϕ_α and represents the kth clock sample for α. The *routing indicator*, $U(\alpha,k)$, is uniformly distributed on $[0,1)$ and will be used to determine the state transition at the kth occurrence of α. State transitions are determined by a mapping $\psi : X \times \Gamma \times [0,1) \to X$, i.e., if α occurs in state x with routing indicator $u = U(\alpha,k)$, the state jumps to $x' = \psi(x,\alpha,u)$. The requirement for ψ is that for all x, α, x'

$$\text{Prob}\left[\psi(x,\alpha,u) = x'\right] = p(x';x,\alpha), \tag{B.1}$$

whenever u is uniformly distributed on $[0,1)$. It is straightforward to show that, with these definitions, the evolution of the trajectory $x(t)$ of the GSMP is governed by

$$\tau_{n+1} = \tau_n + \min\left\{c_n(\alpha) : \alpha \in \Gamma(x_n)\right\}, \tag{B.2}$$

$$a_{n+1} = \arg\left\{\alpha \in \Gamma(x_n) : c_n(\alpha) = \min\left\{c_n(\alpha') : \alpha' \in \Gamma(x_n)\right\}\right\} \tag{B.3}$$

$$N(\alpha,n+1) = \begin{cases} N(\alpha,n)+1, & \alpha = a_{n+1} \\ N(\alpha,n), & \text{otherwise,} \end{cases} \tag{B.4}$$

$$x_{n+1} = \psi\left(x_n, a_{n+1}, U\left(a_{n+1}, N\left(a_{n+1}, n+1\right)\right)\right). \tag{B.5}$$

At each state transition, the clock readings are adjusted by setting clocks for any new events and reducing the time left on any surviving "old"

---

[8] By convention, an event can be scheduled at a_n but does not occur until its clock reading runs down to zero. Thus $N(\alpha,n)$ does not include those α's which have been scheduled but are still alive.

[9] Note, here we assume that the initial clock reading for each occurrence of each event type α is sampled independently from a common distribution ϕ_α. In general, the distribution ϕ_α generating initial reading for a new occurrence of α may depend on the pair of states associated with the occurrence of α and the feasible event set when α occurs.

clocks by the time when the last transition, i.e., the duration of the state x_n or the triggering event's life time. Thus,

$$c_{n+1}(\alpha) = c_n(\alpha) - (\tau_{n+1} - \tau_n), \text{ if } \alpha \in \Gamma(x_n) \text{ and } \alpha \neq a_{n+1}, \quad \text{(B.6)}$$

$$c_{n+1}(\alpha) = Y(\alpha, N(\alpha, n+1)+1), \text{ if } \alpha \in \Gamma(x_{n+1}) \text{ and}$$
$$\text{either } \alpha \notin \Gamma(x_n) \text{ or } \alpha = a_{n+1}. \quad \text{(B.7)}$$

The sample path $x(t)$ is finally generated by setting $x(t) = x_n$ on $[\tau_n, \tau_{n+1})$. $x(t)$ is known as a **Generalized Semi-Markov Process (GSMP)**, and the setup without the timing part with the event sequence given is known as a **Generalized Semi-Markov Scheme (GSMS)**. For mathematical form of the transition kernel of GSMP and a comprehensive treatment of the subject, we refer to (Shedler 1993).

Exercise B.7: When does a GSMP reduce to a continuous time Markov Chain?

Exercise B.8: When does a GSMP reduce to a SMP?

Example: Let us consider a system of two servers connected by a buffer with capacity K as shown in Fig. B.5.

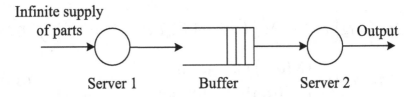

Fig. B.5. A system with two servers and one intermediate buffer

The GSMP model of the system is as follows. We regard the number of parts in the intermediate buffer as the state of the system, x. There are two types of possible events, $\Gamma = \{$ e_1-departure from server 1, e_2-departure from server 2$\}$. Note that, since there is an infinite supply of parts at server 1, we omit the arrival event at server 1. Similarly, when e_1 happens (which presumably requires that the buffer is not full) an arrival event at the buffer immediately happens. So we omit that event, too. Given a state x, the set of feasible events are

$$\Gamma(x) = \begin{cases} \{e_1, e_2\}, & 0 < x < x_{\text{limit}} = K \\ \{e_1\}, & x = 0 \\ \{e_2\}, & x = x_{\text{limit}} = K. \end{cases}$$

The state transition function is

$$x_{\text{next}} = f(x_{\text{now}}, e_{\text{now}}) = \begin{cases} x_{\text{now}} + 1, & \text{if } e_1 \text{ happens} \\ x_{\text{now}} - 1, & \text{if } e_2 \text{ happens and } x_{\text{now}} \neq 0 \\ 0, & \text{if } e_2 \text{ happens and } x_{\text{now}} = 0. \end{cases}$$

The lifetime of the events $c(e_i)$ are samples from exponential distribution with rate μ_i, $i = 1, 2$ (Note breakdown can be considered as an extra long service time).

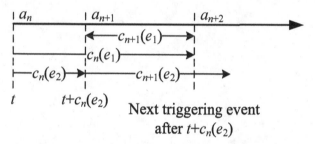

Next triggering event
after $t + c_n(e_2)$

Fig. B.6. Illustration for lifetime update

To simulate the system, the lifetime update process will look like Fig. B.6. Suppose the system just experiences the nth event at time t, both servers are busy, and $x_n = 2$, $K = 3$. So $\Gamma(x_n) = \{e_1, e_2\}$. Suppose the lifetime of both events is as shown in Fig. B.6. Because $c_n(e_2) < c_n(e_1)$, which means server 2 finishes first. Then the system clock moves to $t + c_n(e_2)$. When the $(n+1)$th event $a_{n+1} = e_2$ happens, the state changes $x_{n+1} = f(x_n, e_2) = x_n - 1$. Since event e_1 is not the triggering event and $x_{n+1} < K$, $c_{n+1}(e_1) = c_n(e_1) - c_n(e_2)$. Event e_2 is the triggering event and is still feasible at state x_{n+1}, so a new sample of the lifetime is taken, i.e., $c_{n+1}(e_2)$. Because $c_{n+1}(e_2) > c_{n+1}(e_1)$, we have $a_{n+2} = e_1$, and this procedure continues to simulate the system.

Appendix C Universal Alignment Tables for the Selection Rules in Chapter III

As mentioned in Chapter III, it is practically infeasible to calculate the size of the selected set through a closed form expression. However, there is an exception if we use the Blind Pick rule, and the selected size supplies an upper bound for other selection rules. In Chapter III, we use the following function to approximate the selected size in different scenarios

$$Z(k,g) = e^{Z_1} k^{Z_2} g^{Z_3} + Z_4,$$

where Z_1, Z_2, Z_3, Z_4 are constants depending on the computing budget T, the noise level, the OPC class, and the selection rule thus used. Ample experiments show that these functions are good approximation to the true values (for such an example, please see Section III.2). In Section III.2 and III.3 we show how to compare different selection rules, using the regression functions, and find a good selection rule for the given problems. In this way, we can get a smaller selected set than using a randomly picked selection rule, and further save the computing budget. For the convenience of practical application, we list in this appendix the regressed values of these four coefficients under different scenarios, for all the selection rules considered in Chapter III.

For the BP rule, the regressed values are: $Z_1 = 7.8189$, $Z_2 = 0.6877$, $Z_3 = -0.9550$, and $Z_4 = 0.00$ (Lau and Ho 1997).

For other rules, the regressed values are listed in the following three groups of tables gathered according to the computing budget. Tables C.1-C.8 are for the case when a large computing budget is available; Tables C.9-C.14 are for the medium computing budget; Tables C.15-C.20 are for the small computing budget. In each group, we list the regressed values in the order of HR, OCBA, B vs. D, SPE, HR_gc, HR_ne. The regression values of rule RR and HR_CRR are only listed (at the end of the group) under the large computing budget ($T = 30000$). The reason is that RR and HR_CRR can explore 173 designs under the large computing budget (see Section III.2.1), so we have many choices of selected subset sizes. By contrast, these two selection rules only explore 32 and 22 designs under the medium or small computing budget (see Table 3.3) respectively. To cover

1 or 2 of the top-100 designs with a probability no less than 0.95, we usually need to select all the explored designs, the number of which is 32 and 22 respectively.

Note, since these tables are regressed based on experimental data, these tables are not supposed to be used outside the range of the parameter settings in the experiments in Section III.2.1. To be specific, the working ranges are: $g \in [20, 200]$, $k \in [1, 10]$, $s = Z(\bullet / \bullet) < 180$, and when the fraction k/g is small, in general[1]. For B vs. D, the working range should also satisfy the following conditions:

when $T = 30000$ (the case of large computing budget), $1 \le k \le 5$ and $k/g \le 1/15$;
when $T = 1000$ (the case of medium computing budget), $1 \le k \le 3$ and $k/g \le 1/25$;
when $T = 500$ (the case of small computing budget), $1 \le k \le 2$ and $k/g \le 1/35$.

Table C.1. Regressed values of Z_1, Z_2, Z_3, Z_4 in $Z(k,g)$ for HR when $T = 30000$

Noise			$N(0, 0.25^2)$		
OPC class	Flat	U-shape	Neutral	Bell	Steep
Z_1	7.4106	5.6654	4.1423	2.7231	−5.7774
Z_2	0.7441	1.0882	1.6759	2.4333	5.8835
Z_3	−0.9985	−1.2736	−1.5084	−1.8215	−2.0584
Z_4	4.7407	7.7228	9.3672	9.8303	9.9279
Noise			$N(0, 0.5^2)$		
OPC class	Flat	U-shape	Neutral	Bell	Steep
Z_1	7.5536	6.1168	5.1895	4.1587	0.6777
Z_2	0.6893	0.9970	1.2664	1.6378	3.0335
Z_3	−0.9723	−1.2222	−1.3397	−1.4624	−1.8638
Z_4	0.9345	7.0204	8.7662	9.3611	9.8529
Noise			$N(0, 1.5^2)$		
OPC class	Flat	U-shape	Neutral	Bell	Steep
Z_1	7.8320	6.6726	6.3094	5.9387	4.1500
Z_2	0.7806	0.8670	0.9456	1.2138	2.0904
Z_3	−1.0639	−1.1236	−1.1600	−1.2982	−1.8284
Z_4	5.0703	4.6716	6.6586	8.6564	10.4075

---

[1] Basically we pick the ranges of values to make satisfaction of high AP value relatively easy.

Table C.2. Regressed values of Z_1, Z_2, Z_3, Z_4 in $Z(k,g)$ for OCBA when $T = 30000$

Noise OPC class	$N(0, 0.25^2)$				
	Flat	U-shape	Neutral	Bell	Steep
Z_1	7.8201	7.6707	7.4504	6.5935	−5.0976
Z_2	0.8322	1.9139	3.3586	3.9551	7.3529
Z_3	−1.1179	−2.2932	−3.6225	−4.0905	−3.2942
Z_4	4.3574	9.1222	9.8669	10.0036	9.9414
Noise OPC class	$N(0, 0.5^2)$				
	Flat	U-shape	Neutral	Bell	Steep
Z_1	7.6030	8.1800	8.1473	7.0666	1.0126
Z_2	0.7839	1.6994	3.0569	3.8156	6.6420
Z_3	−1.0374	−2.1591	−3.3355	−3.8050	−4.5920
Z_4	3.5680	8.7907	10.0115	10.0553	9.9868
Noise OPC class	$N(0, 1.5^2)$				
	Flat	U-shape	Neutral	Bell	Steep
Z_1	7.8724	8.7363	9.1567	10.1502	6.7102
Z_2	0.7760	1.5107	2.2478	2.9897	5.2664
Z_3	−1.0598	−1.9826	−2.6426	−3.5029	−4.6962
Z_4	3.8224	8.2605	9.9499	10.6282	10.1121

Table C.2. Regressed values of Z_1, Z_2, Z_3, Z_4 in $Z(k,g)$ for B vs. D when $T = 30000$

Noise OPC class	$N(0, 0.25^2)$				
	Flat	U-shape	Neutral	Bell	Steep
Z_1	7.6177	6.9495	3.4220	−0.3083	−11.1962
Z_2	0.7996	3.3313	9.8183	6.5448	7.8702
Z_3	−0.9838	−2.1278	−4.4306	−2.3014	−1.0668
Z_4	−1.0678	8.4713	10.0513	9.5008	9.7801
Noise OPC class	$N(0, 0.5^2)$				
	Flat	U-shape	Neutral	Bell	Steep
Z_1	7.5211	8.9030	7.1799	2.8575	−8.1332
Z_2	0.7420	3.0512	5.6060	8.4166	9.8915
Z_3	−0.9271	−2.2974	−3.0774	−3.3922	−2.3525
Z_4	−2.6531	8.0662	9.3758	9.4724	10.3882
Noise OPC class	$N(0, 1.5^2)$				
	Flat	U-shape	Neutral	Bell	Steep
Z_1	7.9956	8.7000	8.9830	8.4086	2.1346
Z_2	0.7820	1.8822	2.7129	3.2634	5.6452
Z_3	−1.0457	−1.6177	−1.9892	−2.1224	−1.9475
Z_4	3.8068	3.3109	5.9167	6.4194	8.5570

Table C.4. Regressed values of Z_1, Z_2, Z_3, Z_4 in $Z(k,g)$ for SPE when $T = 30000$

Noise	$N(0, 0.25^2)$				
OPC class	Flat	U-shape	Neutral	Bell	Steep
Z_1	7.7936	6.5980	6.0474	5.2723	2.5343
Z_2	0.8073	1.2619	1.7901	2.0448	3.8335
Z_3	−1.0933	−1.5353	−1.9598	−2.0858	−2.8124
Z_4	4.9850	8.3364	9.7051	9.9103	10.1179
Noise	$N(0, 0.5^2)$				
OPC class	Flat	U-shape	Neutral	Bell	Steep
Z_1	7.6814	6.8916	6.4591	6.0608	3.7260
Z_2	0.7236	1.0760	1.4475	1.7058	2.7574
Z_3	−1.0117	−1.4123	−1.6662	−1.8822	−2.2735
Z_4	2.1583	7.6338	9.2568	9.7934	9.9160
Noise	$N(0, 1.5^2)$				
OPC class	Flat	U-shape	Neutral	Bell	Steep
Z_1	7.9190	7.2864	7.0807	6.8400	6.1009
Z_2	0.8042	1.0390	1.1438	1.2775	2.1433
Z_3	−1.0889	−1.3262	−1.3762	−1.4581	−2.2384
Z_4	7.1896	6.7197	7.2051	8.3168	11.1465

Table C.5. Regressed values of Z_1, Z_2, Z_3, Z_4 in $Z(k,g)$ for HR_gc when $T = 30000$

Noise	$N(0, 0.25^2)$				
OPC class	Flat	U-shape	Neutral	Bell	Steep
Z_1	7.4513	6.2001	4.9004	3.6507	−1.2164
Z_2	0.7225	1.1508	1.7883	2.4183	4.5927
Z_3	−0.9954	−1.4526	−1.8146	−2.0558	−2.5741
Z_4	3.4786	8.0685	9.5995	9.8139	9.9292
Noise	$N(0, 0.5^2)$				
OPC class	Flat	U-shape	Neutral	Bell	Steep
Z_1	7.3884	6.3210	5.7244	4.9559	0.1160
Z_2	0.6630	1.0926	1.3261	1.7836	3.8130
Z_3	−0.9299	−1.3423	−1.5413	−1.7854	−2.1167
Z_4	1.1045	7.4903	8.9882	9.5246	9.8955
Noise	$N(0, 1.5^2)$				
OPC class	Flat	U-shape	Neutral	Bell	Steep
Z_1	7.9538	7.1643	6.7253	6.2830	4.3164
Z_2	0.7779	1.0065	1.2254	1.4287	1.9277
Z_3	−1.0884	−1.3435	−1.4530	−1.5724	−1.8173
Z_4	5.1576	6.6384	8.6297	9.4539	9.7080

Table C.6. Regressed values of Z_1, Z_2, Z_3, Z_4 in $Z(k,g)$ for HR_ne when $T = 30000$

Noise	$N(0, 0.25^2)$				
OPC class	Flat	U-shape	Neutral	Bell	Steep
Z_1	7.4580	5.6590	3.1601	0.1160	15.9239
Z_2	0.7157	1.1175	2.3256	3.8130	7.7238
Z_3	−0.9924	−1.3547	−1.8042	−2.1167	−13.1714
Z_4	1.8543	7.9188	9.6509	9.8955	9.9987
Noise	$N(0, 0.5^2)$				
OPC class	Flat	U-shape	Neutral	Bell	Steep
Z_1	7.5698	6.1261	5.0156	3.5878	−1.2164
Z_2	0.7809	1.1577	1.6711	2.3747	4.5927
Z_3	−1.0366	−1.3887	−1.7040	−1.9490	−2.5741
Z_4	5.0769	7.8866	9.5254	9.7859	9.9292
Noise	$N(0, 1.5^2)$				
OPC class	Flat	U-shape	Neutral	Bell	Steep
Z_1	7.8650	6.8138	6.3370	6.1343	1.2603
Z_2	0.7869	1.0162	1.3027	1.6369	4.0765
Z_3	−1.0745	−1.2953	−1.4677	−1.7640	−2.5663
Z_4	5.0754	6.4476	8.6670	9.8197	10.0970

Table C.7. Regressed values of Z_1, Z_2, Z_3, Z_4 in $Z(k,g)$ for RR when $T = 30000$

Noise	$N(0, 0.25^2)$				
OPC class	Flat	U-shape	Neutral	Bell	Steep
Z_1	7.7227	8.9909	10.8990	15.8410	−20.4900
Z_2	0.8057	1.7490	3.4057	5.6362	135.2500
Z_3	−1.0809	−2.2033	−3.5396	−5.8401	−64.1410
Z_4	5.4032	8.8419	9.5774	9.8350	9.9998
Noise	$N(0, 0.5^2)$				
OPC class	Flat	U-shape	Neutral	Bell	Steep
Z_1	7.8185	8.3533	9.8197	8.4884	41.9990
Z_2	0.8253	1.5218	2.6282	3.3773	13.7690
Z_3	−1.0924	−1.8774	−2.8101	−3.0048	−15.9300
Z_4	6.0076	8.2653	9.4486	9.5267	9.9921
Noise	$N(0, 1.5^2)$				
OPC class	Flat	U-shape	Neutral	Bell	Steep
Z_1	8.1533	9.4407	11.4580	11.8750	7.5157
Z_2	0.8255	1.6098	2.0984	2.3734	5.3458
Z_3	−1.1416	−2.0185	−2.7018	−2.9689	−3.5537
Z_4	6.9006	9.4492	10.4110	10.6130	10.1530

Table C.8. Regressed values of Z_1, Z_2, Z_3, Z_4 in $Z(k,g)$ for HR_CRR when $T = 30000$

Noise	$N(0, 0.25^2)$				
OPC class	Flat	U-shape	Neutral	Bell	Steep
Z_1	7.5996	11.0870	11.0540	121.2000	−0.5696
Z_2	0.7968	2.3842	14.1700	69.0190	1.0714
Z_3	−1.0608	−3.1418	−9.1531	−61.7400	−2.7783
Z_4	4.7842	9.4394	9.9536	9.9998	10.0000
Noise	$N(0, 0.5^2)$				
OPC class	Flat	U-shape	Neutral	Bell	Steep
Z_1	7.4763	9.2477	13.3230	19.8560	121.2000
Z_2	0.7017	2.0832	4.4213	6.9851	69.0190
Z_3	−0.9640	−2.4452	−4.6347	−7.4652	−61.7400
Z_4	0.1384	9.0746	9.7686	9.9103	9.9998
Noise	$N(0, 1.5^2)$				
OPC class	Flat	U-shape	Neutral	Bell	Steep
Z_1	8.0833	9.7982	10.5790	8.5795	6.1287
Z_2	0.8136	1.7538	2.4293	2.7166	6.3867
Z_3	−1.1308	−2.2395	−2.7860	−2.5795	−4.0441
Z_4	5.6291	9.0190	9.7647	9.4184	9.8267

Table C.9. Regressed values of Z_1, Z_2, Z_3, Z_4 in $Z(k,g)$ for HR when $T = 1000$

Noise	$N(0, 0.25^2)$				
OPC class	Flat	U-shape	Neutral	Bell	Steep
Z_1	7.6176	6.6517	6.3875	5.8768	3.0523
Z_2	0.7065	0.8557	1.0385	1.0590	2.5892
Z_3	−0.9745	−1.1390	−1.2502	−1.2228	−1.8703
Z_4	1.5720	5.4693	7.4041	7.8843	10.0340
Noise	$N(0, 0.5^2)$				
OPC class	Flat	U-shape	Neutral	Bell	Steep
Z_1	7.8339	7.0220	6.7000	6.5053	5.1791
Z_2	0.7561	0.8811	0.8582	0.9525	1.3498
Z_3	−1.0340	−1.1356	−1.0580	−1.0813	−1.2239
Z_4	5.4467	4.8870	4.2599	5.7163	9.5050
Noise	$N(0, 1.5^2)$				
OPC class	Flat	U-shape	Neutral	Bell	Steep
Z_1	7.8913	7.7593	7.4124	7.1687	6.2993
Z_2	0.7548	0.8359	0.7818	0.7562	0.7677
Z_3	−1.0220	−1.1369	−1.0182	−0.9477	−0.8030
Z_4	5.2749	4.8319	2.9301	3.8021	2.4132

Table C.10. Regressed values of Z_1, Z_2, Z_3, Z_4 in $Z(k,g)$ for OCBA when $T = 1000$

Noise	$N(0, 0.25^2)$				
OPC class	Flat	U-shape	Neutral	Bell	Steep
Z_1	7.8089	7.4288	8.0146	8.9451	1.5245
Z_2	0.7954	1.2231	1.7613	2.5420	3.1718
Z_3	−1.0697	−1.5286	−2.0759	−2.8765	−1.5684
Z_4	7.5845	7.5827	9.6300	10.6460	9.7590
Noise	$N(0, 0.5^2)$				
OPC class	Flat	U-shape	Neutral	Bell	Steep
Z_1	8.1180	7.8239	7.6229	7.7051	9.4175
Z_2	0.8042	1.1184	1.2125	1.4379	3.6047
Z_3	−1.1244	−1.4416	−1.4811	−1.6493	−3.6548
Z_4	8.2637	7.3700	7.8485	9.1366	11.4610
Noise	$N(0, 1.5^2)$				
OPC class	Flat	U-shape	Neutral	Bell	Steep
Z_1	7.9447	8.1154	7.5361	7.5373	6.4260
Z_2	0.8086	0.9889	0.8426	0.8326	0.9993
Z_3	−1.0640	−1.2872	−1.0786	−1.0689	−0.9580
Z_4	7.4456	8.6345	4.7621	5.3821	6.9053

Table C.11. Regressed values of Z_1, Z_2, Z_3, Z_4 in $Z(k,g)$ for B vs. D when $T = 1000$

Noise	$N(0, 0.25^2)$				
OPC class	Flat	U-shape	Neutral	Bell	Steep
Z_1	7.9996	9.2259	11.2790	9.3277	−14.7780
Z_2	0.7691	2.8732	4.7131	4.3101	13.8290
Z_3	−1.0453	−2.0335	−3.1179	−2.6577	−1.3282
Z_4	2.3862	7.9018	9.1377	8.4729	10.1460
Noise	$N(0, 0.5^2)$				
OPC class	Flat	U-shape	Neutral	Bell	Steep
Z_1	8.6831	9.0142	10.4080	9.6538	3.8676
Z_2	0.9522	1.9785	2.9283	3.3931	9.1748
Z_3	−1.2450	−1.6266	−2.2001	−2.1829	−2.9535
Z_4	8.6772	4.9928	7.4656	7.2247	10.3660
Noise	$N(0, 1.5^2)$				
OPC class	Flat	U-shape	Neutral	Bell	Steep
Z_1	8.3728	7.6015	8.1495	7.1164	7.7334
Z_2	0.8077	1.0014	1.0854	0.6641	1.5640
Z_3	−1.1323	−0.9742	−1.1179	−0.7573	−1.1501
Z_4	6.6783	−4.8144	−1.6719	−16.0820	1.4790

Table C.12. Regressed values of Z_1, Z_2, Z_3, Z_4 in $Z(k,g)$ for SPE when $T = 1000$

Noise	$N(0, 0.25^2)$				
OPC class	Flat	U-shape	Neutral	Bell	Steep
Z_1	7.6637	7.9626	8.0102	7.6565	7.2879
Z_2	0.7208	1.1759	1.4063	1.7317	2.4221
Z_3	−0.9951	−1.5547	−1.7371	−1.9001	−2.5396
Z_4	3.6603	7.5852	8.4178	9.5761	10.5370
Noise	$N(0, 0.5^2)$				
OPC class	Flat	U-shape	Neutral	Bell	Steep
Z_1	8.2896	8.0829	7.9880	7.6714	6.1825
Z_2	0.8558	1.1034	1.1815	1.2806	2.1457
Z_3	−1.1757	1.4399	−1.4802	−1.4449	−1.7480
Z_4	7.8464	6.5954	7.5470	7.0642	10.1390
Noise	$N(0, 1.5^2)$				
OPC class	Flat	U-shape	Neutral	Bell	Steep
Z_1	7.9069	8.1283	7.7164	8.1764	6.7923
Z_2	0.7125	0.8745	0.8020	0.9996	0.9339
Z_3	−1.0032	−1.2053	−1.0612	−1.2505	−0.9515
Z_4	2.7489	4.5238	0.3821	7.2053	4.1635

Table C.13. Regressed values of Z_1, Z_2, Z_3, Z_4 in $Z(k,g)$ for HR_gc when $T = 1000$

Noise	$N(0, 0.25^2)$				
OPC class	Flat	U-shape	Neutral	Bell	Steep
Z_1	7.8084	7.3526	7.5086	7.5179	10.3140
Z_2	0.7724	1.0809	1.4306	1.9640	3.3967
Z_3	−1.0554	−1.4334	−1.7512	−2.1913	−4.3550
Z_4	4.2249	6.8723	8.9380	10.1390	10.2500
Noise	$N(0, 0.5^2)$				
OPC class	Flat	U-shape	Neutral	Bell	Steep
Z_1	8.4852	7.7993	7.8501	8.0894	6.9319
Z_2	0.8782	1.0331	1.2473	1.4704	2.4167
Z_3	−1.2332	−1.4155	−1.5618	−1.7838	−2.3867
Z_4	8.9124	6.4890	7.6601	8.8408	10.6350
Noise	$N(0, 1.5^2)$				
OPC class	Flat	U-shape	Neutral	Bell	Steep
Z_1	8.2202	8.1377	7.9465	7.8395	6.8634
Z_2	0.7957	0.9670	0.9056	0.9906	1.1069
Z_3	−1.1192	−1.2933	−1.1922	−1.2162	−1.1053
Z_4	5.9654	5.2711	2.8148	2.3602	5.5796

Table C.14. Regressed values of Z_1, Z_2, Z_3, Z_4 in $Z(k,g)$ for HR_ne when $T = 1000$

Noise OPC class	$N(0, 0.25^2)$				
	Flat	U-shape	Neutral	Bell	Steep
Z_1	8.0989	7.4390	7.7019	8.2978	8.6701
Z_2	0.8360	1.2591	1.6555	2.4837	4.4201
Z_3	−1.1593	−1.5959	−2.0192	−2.8377	−4.6661
Z_4	8.3361	8.2139	9.7230	10.5130	10.0050
Noise OPC class	$N(0, 0.5^2)$				
	Flat	U-shape	Neutral	Bell	Steep
Z_1	8.1978	7.8594	8.0260	8.5885	10.2590
Z_2	0.8354	1.1403	1.4579	1.7968	4.5748
Z_3	−1.1576	−1.5133	−1.7708	−2.1707	−4.8329
Z_4	8.0641	7.2528	9.0181	10.2110	10.9850
Noise OPC class	$N(0, 1.5^2)$				
	Flat	U-shape	Neutral	Bell	Steep
Z_1	8.2083	8.2428	8.5056	8.0884	6.9772
Z_2	0.8153	1.0505	1.1461	1.0862	1.3409
Z_3	−1.1295	−1.3622	−1.4481	−1.3248	−1.2707
Z_4	8.2221	5.1896	6.5376	5.4885	6.7126

Table C.15. Regressed values of Z_1, Z_2, Z_3, Z_4 in $Z(k,g)$ for HR when $T = 500$

Noise OPC class	$N(0, 0.25^2)$				
	Flat	U-shape	Neutral	Bell	Steep
Z_1	7.7731	7.1650	7.3319	7.1950	4.0657
Z_2	0.7264	0.9531	1.1572	1.3901	2.3351
Z_3	−1.0167	−1.2756	−1.4785	−1.6428	−1.7081
Z_4	2.4674	6.2832	7.9417	9.5867	9.7219
Noise OPC class	$N(0, 0.5^2)$				
	Flat	U-shape	Neutral	Bell	Steep
Z_1	7.9105	7.5306	7.2710	7.3699	6.6832
Z_2	0.8111	0.9630	0.9849	1.0587	1.4206
Z_3	−1.0844	−1.2541	−1.2284	−1.2850	−1.4840
Z_4	8.4549	5.6330	6.8210	6.9921	9.9519
Noise OPC class	$N(0, 1.5^2)$				
	Flat	U-shape	Neutral	Bell	Steep
Z_1	7.9382	7.9455	7.6651	7.3648	6.9159
Z_2	0.7637	0.8855	0.7993	0.7934	0.9468
Z_3	−1.0371	−1.1706	−1.0639	−0.9880	−0.9938
Z_4	5.3905	5.4909	3.9763	3.6898	6.6152

Table C.16. Regressed values of Z_1, Z_2, Z_3, Z_4 in $Z(k,g)$ for OCBA when $T = 500$

Noise OPC class	$N(0, 0.25^2)$				
	Flat	U-shape	Neutral	Bell	Steep
Z_1	7.7380	7.3466	7.1883	7.7230	1.8935
Z_2	0.7621	1.0370	1.2568	1.6020	2.9032
Z_3	−1.0337	−1.3492	−1.4843	−1.8595	−1.4038
Z_4	5.9884	7.1513	8.8643	10.5810	9.4089
Noise OPC class	$N(0, 0.5^2)$				
	Flat	U-shape	Neutral	Bell	Steep
Z_1	7.7007	7.3191	7.2470	7.1794	5.4955
Z_2	0.7047	0.9004	0.9521	1.0204	1.6730
Z_3	−0.9813	−1.1682	−1.1806	−1.1988	−1.2769
Z_4	3.9332	5.6401	7.0346	8.0729	9.5205
Noise OPC class	$N(0, 1.5^2)$				
	Flat	U-shape	Neutral	Bell	Steep
Z_1	8.0856	7.9313	7.4405	7.8183	6.7244
Z_2	0.7670	0.8540	0.7909	0.8536	0.8056
Z_3	−1.0649	−1.1505	−1.0083	−1.1151	−0.8763
Z_4	5.7885	7.1222	4.4384	8.4226	4.5085

Table C.17. Regressed values of Z_1, Z_2, Z_3, Z_4 in $Z(k,g)$ for B vs. D when $T = 500$

Noise OPC class	$N(0, 0.25^2)$				
	Flat	U-shape	Neutral	Bell	Steep
Z_1	7.3718	13.5319	15.9734	16.5227	−9.7923
Z_2	0.7000	2.9687	5.7224	7.7479	10.0888
Z_3	−0.8711	−2.9659	-4.1205	−4.8459	−0.5809
Z_4	−2.4103	9.0682	9.2433	10.0123	9.8663
Noise OPC class	$N(0, 0.5^2)$				
	Flat	U-shape	Neutral	Bell	Steep
Z_1	8.0476	11.0422	10.4414	13.5566	12.8275
Z_2	0.7000	2.1440	2.1090	3.4419	5.1558
Z_3	−1.0513	−2.0570	−1.8891	−2.8171	−3.3038
Z_4	4.5822	7.0766	5.6979	7.8229	9.6202
Noise OPC class	$N(0, 1.5^2)$				
	Flat	U-shape	Neutral	Bell	Steep
Z_1	7.3680	8.7349	8.3968	8.7761	7.7309
Z_2	0.5359	1.0300	1.0355	1.1269	0.9650
Z_3	−0.8452	−1.2109	−1.1165	−1.2499	−0.9926
Z_4	0.3320	−1.2333	−2.7465	3.2621	−0.8444

Table C.18. Regressed values of Z_1, Z_2, Z_3, Z_4 in $Z(k,g)$ for SPE when $T = 500$

Noise	$N(0, 0.25^2)$				
OPC class	Flat	U-shape	Neutral	Bell	Steep
Z_1	8.2899	9.7303	10.2214	10.7590	15.0883
Z_2	0.8954	1.4680	1.7764	2.2176	4.1650
Z_3	−1.2004	−1.9897	−2.2626	−2.6307	−4.8526
Z_4	9.6685	8.6530	9.3290	10.1485	11.9774
Noise	$N(0, 0.5^2)$				
OPC class	Flat	U-shape	Neutral	Bell	Steep
Z_1	8.2139	9.7071	8.7965	9.0619	8.2717
Z_2	0.8816	1.3031	1.1979	1.3334	1.9307
Z_3	−1.1656	−1.7979	−1.5557	−1.6662	−1.8624
Z_4	10.1190	7.5504	5.7980	6.6080	8.6316
Noise	$N(0, 1.5^2)$				
OPC class	Flat	U-shape	Neutral	Bell	Steep
Z_1	8.1022	8.5482	8.2760	8.2313	7.7431
Z_2	0.7745	0.9678	0.8482	0.8546	0.9573
Z_3	−1.0727	−1.3065	−1.1730	−1.1569	−1.1163
Z_4	6.0978	6.9712	3.3010	2.0780	3.3132

Table C.19. Regressed values of Z_1, Z_2, Z_3, Z_4 in $Z(k,g)$ for HR_gc when $T = 500$

Noise	$N(0, 0.25^2)$				
OPC class	Flat	U-shape	Neutral	Bell	Steep
Z_1	7.9373	9.5833	9.9504	10.7244	19.6041
Z_2	0.8064	1.5247	2.0581	2.4866	6.7994
Z_3	−1.0845	−2.0869	−2.4716	−2.9426	−7.6657
Z_4	5.1087	9.6054	10.2930	10.3372	10.8282
Noise	$N(0, 0.5^2)$				
OPC class	Flat	U-shape	Neutral	Bell	Steep
Z_1	8.2384	9.4548	9.5925	9.0835	10.6036
Z_2	0.8883	1.3949	1.4859	1.6741	3.2771
Z_3	−1.1736	−1.8758	−1.9621	−1.9369	−3.2191
Z_4	8.3728	8.6266	8.8970	8.4971	11.0202
Noise	$N(0, 1.5^2)$				
OPC class	Flat	U-shape	Neutral	Bell	Steep
Z_1	8.0678	8.9449	8.7726	8.2823	7.8937
Z_2	0.7799	1.0792	1.0460	0.9497	1.0951
Z_3	−1.0733	−1.4492	−1.3876	−1.2438	−1.2442
Z_4	6.1751	5.3848	5.6418	5.1152	4.6148

Table C.20. Regressed values of Z_1, Z_2, Z_3, Z_4 in $Z(k,g)$ for HR_ne when $T = 500$

Noise	$N(0, 0.25^2)$				
OPC class	Flat	U-shape	Neutral	Bell	Steep
Z_1	8.0214	9.0680	10.3896	13.3639	5.7683
Z_2	0.8184	1.6085	2.3468	3.2543	5.4417
Z_3	−1.1082	−2.0533	−2.7877	−4.0663	−3.6601
Z_4	5.8159	9.2108	10.1850	10.7751	9.8589
Noise	$N(0, 0.5^2)$				
OPC class	Flat	U-shape	Neutral	Bell	Steep
Z_1	8.4357	9.9937	10.0117	10.4979	13.6679
Z_2	0.8980	1.5061	1.6846	2.2172	5.6006
Z_3	−1.2208	−2.0740	−2.1761	−2.5463	−5.2272
Z_4	8.8217	9.2039	8.9303	9.7228	11.1670
Noise	$N(0, 1.5^2)$				
OPC class	Flat	U-shape	Neutral	Bell	Steep
Z_1	8.1816	8.8590	8.7609	8.5659	8.0924
Z_2	0.8103	1.0588	1.1191	1.0818	1.2562
Z_3	−1.1138	−1.4320	−1.4248	−1.3664	−1.3644
Z_4	7.7265	4.6969	5.4774	5.6254	4.9937

Appendix D Exercises

This appendix contains some additional exercises for readers to test their understanding of the previous chapters as well as topics in general discrete event system which are usually supplemental materials in a course using this book as the main text. Problems which are related to but require knowledge outside of the content of this book are indicated by a "*". These give some indication of additional materials typically covered in our courses at Harvard and Tsinghua University. Three groups of exercises are presented. First, the True/False group, in which the answers are simply binary choices. Second, the multiple-choice group, in which the answer is one of the several candidates. Third, the general group, in which the answer may be one or two sentences, or may need some calculation.

1 True/False questions

1. Discrete event systems and discrete time systems (governed by difference equations) both have piecewise constant trajectories. These two names do not denote the same class of systems and are not interchangeable. (T or F)
2*. In a closed queuing network with an unlimited queue size, perturbations in start/stop times of a server can be propagated from one server to another only through the start of busy periods of a server. (T or F)
3. If I want to generate samples from a Gaussian distribution with a given mean and variance, I can always derive from them samples for an arbitrary uniform distribution. (T or F)
4. In the simulation of stochastic systems, the confidence interval of an estimate can decrease faster than $1/(t)^{1/2}$ if there are correlation among the random variables used in the simulation. (T or F)
5. In the generation of random numbers on (0,1) using the linear congruential method, it is very important to choose the coefficients a and b in the formula $x_{n+1} = \text{mod}_m[ax_n+b]$, in order to form a pseudo random sequence. (T or F)
6. It is not important to choose the random seed x_0 in the above question. (T or F)

7. We get one sample path of a DEDS through simulation and compute from this sample path a sample performance for the DEDS. Call this $L(\theta,\xi)$ where θ are the system parameters and ξ the random realizations of various random variables. We are interested in $J = E[L(\theta,\xi)]$. Then a single sample of $L(\theta,\xi)$ is an unbiased estimate of J. (T or F)

8. In the above question, not knowing anything else, would you say L is a good estimate of J? (Y or N)

9. If the expected performance J is the waiting time of customers in a G/G/1 queue, and the sample path and the simulation go forward in time, then L, defined to be the average of the waiting times of all past customers, will become an increasing good estimate of J. (T or F)

10*. Alternatively, if we simulate the G/G/1 queue only for 10 customers per simulation run but we average over many different short simulation runs, each starting from the same initial condition (different random seed of course), then the average of these transient customer waiting times will not approach J, as the number of simulation runs approaches infinity. (T or F)

11*. In the GSMP model of a DEDS, the propagation rule of perturbation analysis is universal. (T or F)

12*. For an n-node queuing network in steady state, the fundamental quantity of interest is the probability distribution of the state of the network, $P(x_1, \ldots, x_n)$, where x_i is the number of customers at node i. From this all important performances can be calculated. (T or F)

13. Ordinal optimization, like other soft computing tools, is heuristic and not easily quantifiable. (T or F)

14. In the GSMP model of DEDS, the specification of the clock mechanism is independent of that of the state transition function. (T or F)

15. In the Standard Clock (SC) scheme of DEDS simulation, the event timing and event typing are totally independent of each other. (T or F)

16. The SC method is independent of the type of distribution of the randomness involved. (T or F)

17*. The so-called traffic equation governing the mean arrival rate at each node of the queuing network is totally general and independent of the distribution type of the service time at each node. (T or F)

18. Sum of Poisson arrivals are still Poisson. (T or F)

19. We can change the mean of a Poisson stream by simply re-scaling the time axis. (T or F)

20. We can also reduce the mean of a Poisson process by simply dropping out each event of a Poisson stream, according to some independent probabilistic means, such as tossing a biased coin. (T or F)

21. The events dropped out of the above question happen to form another Poisson stream. (T or F)

22. In learning literature, the term "generalization" means that the learned model is used to predict the behavior of those unseen examples. Since the generalization performance depends on how well our model can extract information from the training set, a more sophisticated model (model with more features) will always outperform rough models. (T or F)

23*. In an M/M/1 queue, the number of customers served in one busy period is independent of the number of customers served in a different busy period. (T or F)

24. In general, it is easier to determine A>B or A<B than to determine $A-B$=?

2 Multiple-choice questions

25. Consider a regional electric power system with its many generators, transmission lines, various rules for scheduling generation and distribution of power to satisfy demands subject to environmental conditions such as weather. Should this be a DEDS, CVDS or a Hybrid system? Please pick one.

26. Under i.i.d. sampling, the confidence interval of the sample mean as an estimate of the true mean decreases as
    a) $1/n$, where n is the number of samples taken,
    b) $1/(n)^{1/2}$,
    c) $1/(n)^2$,
    d) $a-bn$, where a and b are constants depending on the problem.

27. Suppose you randomly take 1000 samples from an arbitrary distribution and order these samples. The probability that none of the observed samples belongs to the top 1% of the underlying distribution is
    a) absolutely zero,
    b) $1-(1-0.01)^{1000}$
    c) $(1-0.01)^{1000}$
    d) involving summing over a series with many terms too complicated to write down here.

28. In terms of ordinal optimization in the above problem, what is the "good enough" set, G, and what is the "selected" set, S, assuming we are maximizing.
    a) G = top 1% of the distribution and S = the 1000 samples
    b) G = the 1000 samples and S = top 1% of the distribution
    c) G = the largest value of the 1000 samples and S= the largest value of the distribution
    d) G= top 1% of the distribution and S= top 1% of the 1000 samples

29. The probability we are calculating in Exercise 27 is called the "alignment probability" in ordinal optimization.
    a) true,
    b) false
30. The alignment probability approaches one exponentially fast as we increase the size of G and S.
    a) true,
    b) false
31. In OO, the existence of a nonzero mean in the observation noise/error
    a) does not affect the alignment probability,
    b) does not affect the alignment probability if the noises/errors have identical nonzero mean,
    c) affects the alignment probability even if the noises/errors have identical nonzero mean.
32. A pure Markovian DEDS system means
    a) state transition matrix is constant,
    b) state transition time is exponentially distributed at a constant rate,
    c) system is stable,
    d) none of the above,
    e) all of the above.
33*. Little's Law is a relationship between the average number of customers in a system, the average throughput of the system, and the average transit time through the system. It is applicable to
    a) any system in steady state,
    b) only systems that can be described as a queuing network,
    c) only systems in which the randomness is characterized by the Markov assumption,
    d) none of the above.
34*. What the No-Free-Lunch Theorem states is that
    a) estimation of performance value has a theoretical limit of $1/(N)^{1/2}$,
    b) estimation of ordinal information converges at an exponential rate,
    c) averaged over all possible search algorithms, any problem is as hard as any others,
    d) averaged over all possible problems, any search algorithm will be equally efficient,
    e) none of the above.
35. If we have N features, each of which has V different possible values, what is the total cardinality of our feature space?
    a) V^N
    b) N^V
    c) $2^{N/V}$
    d) 2^{NV}.

36. Which of the following statement is NOT true?
    a) Causation implies correlation but correlation does not imply causation.
    b) If A and B have a common cause, the change of A brings the change of B in general.
    c) If variable A causes variable B, the change of A brings the change of B in general.
    d) Two variables may be correlated because they have a common cause.
37. Given the following Bayesian network (in Fig. D.1), which is the correct expression of the joint distribution $P(A, B, C, D, E)$

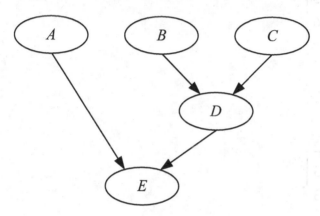

Fig. D.1. A Bayesian network

    a) $P(A)P(B/D)P(C/D)P(D/E)$
    b) $P(A)P(B)P(C)P(D)P(E)$
    c) $P(A/E)P(D/E)P(B/D)P(C/D)$
    d) $P(A)P(B)P(C)P(D/B,C)P(E/A, D)$.
38. Let X be a uniform distributed random variable on $(0,1]$. Which of the following defines an exponentially distributed random variable with mean μ?
    a) $\mu\ln(X)$
    b) $-\ln(X)$
    c) $-\mu\ln(1-X)$
    d) $-(1/\mu)\ln(1-X)$
    e) $-\mu\ln(1+X)$
39. Given a 3-state stochastic finite state machine (equivalently an ergodic Markov Chain) with an equal transition probability from any state to

any other state, what is the steady-state distribution of the system when

the initial state of system is $x_0 = \begin{pmatrix} 1 \\ 0 \\ 0 \end{pmatrix}$?

a) $\begin{pmatrix} 1/2 \\ 1/4 \\ 1/4 \end{pmatrix}$

b) $\begin{pmatrix} 2/3 \\ 1/6 \\ 1/6 \end{pmatrix}$

c) $\begin{pmatrix} 1 \\ 1 \\ 1 \end{pmatrix}$

d) $\begin{pmatrix} 1/2 \\ 1/3 \\ 1/3 \end{pmatrix}$

e) $\begin{pmatrix} 1/3 \\ 1/3 \\ 1/3 \end{pmatrix}$.

40. In a room of 10 people, what is the probability of having at least two people born in the first 36 days of the year (for simplicity, assume a 360 day calendar year)?
    a) $1-(0.9)^{10}$
    b) $(0.9)^{10}$
    c) $1-[(0.9)^{10}+(0.1)(0.9)^9]$
    d) $[(0.9)^{10}+(0.9)^9]$
    e) $1-[(0.9)^{10}+(0.9)^9]$.

41*. Assume that you are an out-of-town visitor who does not know the subway schedule at Harvard Square (which is just outside the gate of Harvard University). The train arrives every 10 minutes at the Harvard station. If you walk into the Harvard subway station at random, how long should you expect to wait to get on the train?
    a) 5 minutes
    b) 3 minutes

    c) 1 minute

    d) 10 minutes.

42. Which of the following is NOT absolutely needed as an ingredient for a DEDS model?

    a) A state transition function,

    b) A feasible event list as a function of the state,

    c) A random number generator,

    d) A collection of possible states.

43. The Birthday Paradox states that, in a room of 23 persons, there is at least as big a chance to get a "head" in flipping a fair coin as for any two persons to have the same birthday. In one of the Tonight Show (America's most popular late night talk shows) in 70s, the former host Johnny Carson attempted to demonstrate the Birthday Paradox by checking his audience (which was nearly 120 in number) to see if anyone had the same birthday as his, say, March 12. He tried three times (i.e., three different individuals), but none of them even had their birthdays in March. Which of the following is true?

    a) Mr. Carson would be very likely to find some matching birthdays if he had surveyed the birthday information of his audience.

    b) The Birthday Paradox does not hold for more than 23 persons in the room.

    c) Mr. Carson should have tried all of his audience and he would have 0.5 chance of hitting someone of his own birthday.

    d) The Birthday Paradox would have been true if Mr. Carson had tried the experiment every day on his show during 1970s.

3 General questions

44. Compare the state of DEDS and CVDS. What's the difference? Try to give an example of each system and enumerate the difference of the states. [Hint: We want you to summarize the difference between DEDS and CVDS.]

45*. There are three queuing systems shown in Fig. D.2. The inter-arrival time and the service time are in exponential distribution at rate λ and μ. The question is which system has the shortest average waiting time?

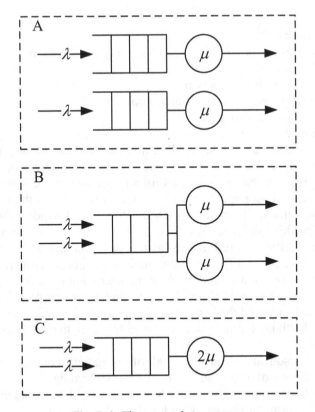

Fig. D.2. Three queuing systems

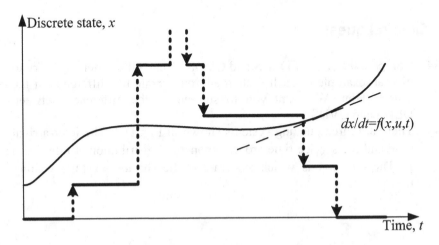

Fig. D.3. The trajectories of a DEDS and a CVDS, where *u* is the input

46. Suppose we integrate the DEDS trajectory in Fig. D.3 twice, the piece wise constant trajectory will look rather smooth and similar to the CVDS trajectory on the same figure. Perhaps we can employ the more familiar CVDS control theory techniques to deal with DEDS. Why don't we do that?

47. With all other parameters remaining the same, in what sense is the Burger King server (one giant queue with multiple servers) better than Macdonald server (one separate queue for each server)? What assumptions are required to justify your answer?

48. Using the definition of a random (stochastic) sequence as a collection of random variables in a general joint distribution $p(x_1, \ldots, x_n)$, define mathematically the following (using one line of mathematical formula or one sentence description. We are looking for conceptual understanding.)
   a) Markov random sequence
   b) Gaussian random sequence
   c) Semi-Markov sequence
   d) Renewal sequence.

49. What is the twin evil of nonlinearity and noise with respect to the formula below?

$$\frac{dE[L]}{d\theta} \approx \frac{\frac{1}{N}\sum_{i=1}^{N} L(\theta + \Delta\theta, \xi_i) - \frac{1}{N}\sum_{i=1}^{N} L(\theta, \xi_i)}{\Delta\theta}.$$

50. If I flip a biased coin (probability of head 3/4) and generate a sequence of heads and tails, I can produce from this sequence another sequence which is statistically indistinguishable from one that could have been generated from a fair coin. What do you need in order to do this?

51. Blind pick alignment probability problem
   If one blindly picks $|S|$ choices out of N possibilities, what is the probability that at least $k<|S|$ of the truly top-$|G|$ choices will be contained in the set S? [Hint: First consider the special case $|S|=|G|$ and then generalize to $|S|\neq|G|$.]

52. A Life-or Death Decision Problem
   Suppose you are facing a crucial decision problem, say a serious medical life-or-death decision involving cancer treatment, or a lifetime marriage partner choice, between courses of action A and B. By spending one million dollars under A, you will find the best decision for sure vs. a cost of $1Million/x dollars under B to get a decision on which is guaranteed to be within the top-5% of all decision choices with a probability equal to 0.99. At what value of x will you be indifferent between the two courses of action?

53. A Wrong Classification Problem

Intuitively, we all know that the longer we run the simulation, the smaller the uncertainty associated with an estimate of the performance J of the system is. Suppose you are told that the uncertainty associated with J is uniformly distributed. The estimate is of half width $w(t)=c/t$, where c is a constant associated with the simulation. Now consider two parametrically different but structurally similar simulation running side-by-side. Their uncertainty constants are c_1 and c_2 respectively. Let the estimates for J be $\hat{J}_1(t)$ and $\hat{J}_2(t)$ respectively at time t. (Fig. D.4 shows one possible situation.) I would like to stop the simulation at a time when I observe, say, $\hat{J}_1(t) > \hat{J}_2(t)$, and I can be 90% sure that the actual order is also $J_1 > J_2$ but not $J_2 > J_1$. It would be the same if the other way around.

(i) To simplify the matters, let us pose an easier problem. Suppose at some time we observe $\hat{J}_1(t) = 1$ and $\hat{J}_2(t) = 2$ and we also know, from statistical analysis, that $w_1 = w_2 = c_1/t = c_2/t = 1.0$. What is the probability that the actual order is $J_1 > J_2$, i.e., the probability of a wrong classification?

(ii) Now suppose that you know c_1 and c_2 and you observe $\hat{J}_1(t)$ and $\hat{J}_2(t)$, describe a procedure (as the function of time) based on which you can be 90% sure that the observe order is in fact the correct order (Describe only the main idea, you are not required to work out the details). [Hint: consider the probability of wrong classification and the diagram below.]

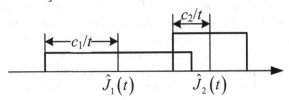

Fig. D.4. Possibilities of wrong classification

(iii) Is the above figure the only possibility you should consider for part (ii)? If so or if not, explain your answer.

(iv) Suppose we observe $\hat{J}_1(t) = 1$ and $\hat{J}_2(t) = 2.5$ at a certain time and we know that $w_1 = c_1/t = 5.0$ and $w_2 = c_2/t = 1.0$ from statistical analysis. What is the probability that the actual order is $J_1 > J_2$, i.e., the probability of a wrong classification? [Hint: consider the answer to part (iii).]

54. Seeing is Believing
Suppose A and B are two random variables with mean "a" and "b", $a>b$. The observation noise $w_1=A-a$ and $w_2=B-b$ are i.i.d. distributed. Prove that

$$\text{Prob}[A>B] \geq \text{Prob}[A<B],$$

i.e., what is observed to be greater is more likely to be actually greater, or the observed order be the actual order.

55. Bounding Blind Pick
Suppose you sample independently from a large (infinite) population of design choices. Each sample, i, represents a particular design resulting in a particular system performance, J_i. Let us say the J_i's are distributed $N(\mu, \sigma^2)$. Now we randomly pick k designs J_i, $i=1, 2, \ldots, k$. What is the probability that among these k picked designs there is at least one that is larger than some given value $J_{\text{satisfactory}}$?

56. More Blind Pick
Consider the same question as above, except that now the population is finite ($i=1, 2, \ldots, N$) with $J_i=i$. We must observe J_i through $N(0,1)$ additive noise, i.e.,

$$J_{\text{observed}} = J_{\text{actual}} + w, \; w \sim N(0,1).$$

(i) What is the probability that it will turn out to be indeed the true top value if we choose the observed top value out of the population?
(ii) What is the probability that at least one will turn out to be among the actual top-2 values if we choose the 2 observed top values?
(iii) Can you generalize the above?
If it helps to simplify things, you can choose a specific value for N, say $N=3$ or 5.

57. A five minute demonstration of OO
Implement a spread sheet example of ordinal optimization as described in Section II.3 (also repeated as follows).

Design # = θ	True performance $J(\theta)$	Noise $w \in U[0,W]$	Observed performance $J(\theta)+w$
1	1.00	87.98	88.98
2	2.00	1.67	3.67
.	.	.	.
.	.	.	.
.	.	.	.
199	199.00	32.92	231.92
200	200.00	24.96	224.96

Sort on this column
in ascending order

Fig. D.5. Spread sheet implementation of generic experiment

In Fig. D. 5, column 1 models the N (=200) alternatives and the true order 1 through N. Column 2 shows the linearly increasing OPC from 1 to N (=200). The noise variance σ^2 essentially determines the estimation or approximation error of $J(\theta)$. This is shown by the random noise generated in column 3, which has a large range $U(0,100)$ in this case. Column 4 displays the corrupted (or estimated) performance. When we sort on column 4 we can directly observe the alignment in column 1, i.e., how many numbers 1 through g (=12) are in the top-g rows. Try this and you will be surprised! It takes less than two minutes to setup on Excel. Replicate the spread sheet calculations several times to get an estimate of $\text{Prob}[|G \cap S| \geq k]$=?

58. It is well known that the accuracy of a sampling scheme depends only on the size of the sample but NOT the size of the underlying population which we shall assume to be infinite in this problem. Suppose the distribution of performances J of a system is normal $N(0, \sigma_J^2)$ when plotted against a design variable θ. We randomly sample N designs, $\theta_1, \theta_2, ..., \theta_N$ in an experiment and observe the system performance, $J(\theta_i)$ in additive noise $N(0,\sigma_n^2)$, i.e., $J(\theta_i)_{\text{observed}} = J(\theta_i)$+noise. We now ask the question what is $\text{Prob}[\max(J(\theta_i)_{\text{observed}}: i=1,...N) \in \text{top } 5\%$ of $J(\theta)]\equiv p$=?, assuming we are interested in maximizing performance. (i) Purely as a test of your probabilistic intuition what do you think is the likely value of p for the case $\sigma_J^2=\sigma_n^2$ and $N=100$?

    a) $p \leq 0.4$,   b) $0.4 < p \leq 0.75$,   c) $p > 0.75$.

    Choose one alternative and calculate the answer to see if you are correct. (ii) Whatever the value of p in (i) is, now consider doing m independent experiments of N samples each and ask $\text{Prob}[$at least one of the $\max(J(\theta_i)_{\text{observed}}: i=1,...N)$ of the m experiments$\in$top 5% of $J(\theta)]\equiv p(m)$=? Calculate $p(m)$ as a function of m and p. What did you learn from this calculation? (Try out a couple of your guesses in (i) for different values of m.) (iii) Suppose, instead of getting m, we did ONE experiment with mN samples and asked $\text{Prob}[\max(J(\theta)_{\text{observed}}: i=1,...mN) \in \text{top } 5\%$ of $J(\theta)]\equiv p^*$=? Is $p^* > p(m)$ or $p(m) > p^*$? Can you relate this to the idea of ordinal optimization?

59. There are ten designs, the performances of which are non-identical and unknown. The designs are labeled as class A and B randomly so that 1) there are 5 A's and 5 B's; 2) the probability that either the top 1 or the top 2 design being contained in class A is 3/4. Each time, we can observe all the labels of the designs, but not the performances, thus the designs within the same class are indistinguishable. The labels of the designs may change from observation. If the good

enough set G is defined as the top-2 designs, and we plan to select 2 balls, i.e., selected subset $|S|=2$.

(i) What is the alignment probability $\text{Prob}[|G \cap S| \geq 1]$, if S is selected by blind pick ?

(ii) If you want to maximize the alignment probability, how do you select your S (explain why), and what is the alignment probability of this way of picking S?

60. This is a problem of determining how "representative" is 1000 samples in capturing an arbitrary distribution.

(i) What is the probability that none of the 1000 samples is contained in the top-1% of the population?

(ii) What is the probability that at least $n(\leq 20)$ of the 1000 samples are contained in the top-1% of the population?

References

Aarts EHL, Korst J (1989) Simulated annealing and Boltzmann machines: A stochastic approach to combinatorial optimization and neural computing. John Wiley and Sons Inc, Chichester England

Andersen HR (1997) An introduction to binary decision diagrams. http://sourceforge.net/projects/buddy

Balakrishnan N, Cohen AC (1991) Order statistics and inference. Academic, New York

Banal R, Basar T (1987) Stochastic teams with nonclassical information revisited: When is an affine law optimal. IEEE Transactions on Automatic Control 32:554–559

Bertsekas DP, Tsitsiklis JN (1996) Neuro-dynamic programming. Athena Scientific, Belmont, MA

Bloch A (1991) The complete Murphy's law. Price Stern Sloan Publisher, Los Angeles, CA

Bo TZ, Hammond JH, Abernathy, FH (1994) Design and scheduling of apparel manufacturing systems with both slow and quick production lines. In: Proceedings of the 33rd IEEE Conference on Decision and Control, pp 1603–1608

Bollig B, Wegener I (1996) Improving the variable ordering of OBDDs is NP-complete. IEEE Transactions on Computers 45(9):993–1002

Bouhia S (2004) A risk-based approach to scheduling and ordering production in manufacturing systems: Real options on production capacity. Ph.D. thesis, Harvard University

Bratley P, Fox BL, Schrage LE (1987) A guide to simulation, 2nd edn. Springer-Verlag, New York

Bryson AE, Ho YC (1979) Applied optimal control. John Wiley and Sons Inc, Chichester England

Cao XR (2005) Basic ideas for event-based optimization of Markov systems. Discrete Event Dynamic Systems: Theory and Applications 15:169–197

Cao XR, Ren ZY, Bhatnagar S, Fu M, Marcus S (2002) A time aggregation approach to Markov decision processes. Automatica 38(6):929–943

Carlier J (1978) Ordonnancements a constraintes disjonctives. RAIRO Recherche operationelle/Operations Research 12:333–351

Cassandras CG (1993) Discrete event systems: Modeling and performance analysis. Richard D Irwin Inc, Homewood, IL

Cassandras CG, Lafortune S (1999) Discrete event systems. Kluwer Academic Publishers, Norwell, MA

Cassandras CG, Strickland SG (1989) On-line sensitivity analysis of Markov chains. IEEE Transactions on Automatic Control 34:76–86

Cassandras CG, Dai L, Panayiotou CG (1998) Ordinal optimization for a class of deterministic and stochastic discrete resource allocation problems. IEEE Transactions on Automatic Control 43(7):881–900

Chambers L (ed) (1995) Practical handbook of genetic algorithms: New frontiers, vol 2. CRC Press, Boca Raton, FL

Chambers L (ed) (1999) Practical handbook of genetic algorithms: Complex coding systems, vol 3. CRC Press, Boca Raton, FL

Chambers L (ed) (2000) The practical handbook of genetic algorithms: Applications, vol 1. Chapman and Hall/CRC Press, Boca Raton, FL

Chen CH (1994) An efficient approach for discrete event system decision problems. Ph.D. thesis, Harvard University

Chen CH, Ho YC (1995) An approximation approach of the standard clock method for general discrete-event simulation. IEEE Transactions on Control Systems Technology 3(3):309–317

Chen CH, Wu SD, Dai L (1997) Algorithm comparison for manufacturing scheduling problems. In: Proceedings of the 36th IEEE Conference on Decision and Control, vol 2, pp 1214–1215

Chen CH, Lin J, Yücesan E, Chick SE (2000) Simulation budget allocation for further enhancing the efficiency of ordinal optimization. Discrete Event Dynamic Systems: Theory and Applications 10(3):251–270

Chen CH, Donohue K, Yücesan E, Lin JW (2003) Optimal computing budget allocation for Monte Carlo simulation with application to product design. Simulation Modelling Practice and Theory 11(1):57–74

Chen EJ (2004) Using ordinal optimization approach to improve efficiency of selection procedures. Discrete Event Dynamic Systems: Theory and Applications 14:153–170

Chen L, Narendra KS (2001) Nonlinear adaptive control using neural networks and multiple models. Automatica, Special issue on neural network feedback control 37(8):1245–1255

Çinlar E (1975) Introduction to stochastic processes. Prentice Hall Inc, Englewood Cliffs, NJ

Coello CAC (2000) An updated survey of GA-based multiobjective optimization techniques. ACM Computing Surveys 32:109–143

Collins NE, Eglese RW, Golden BL (1988) Simulated annealing – an annotated bibliography. American Journal of Mathematical and Management Sciences 8(3):209–307

Croce FD, Tadei R, Volta G (1995) A genetic algorithm for the job shop problem. Computers and Operations Research 22(1):15–24

Dai L (1996) Convergence properties of ordinal comparison in simulation of discrete event dynamic systems. Journal of Optimization Theory and Applications 91(2):363–388

Dai L, Chen CH (1997) Rates of convergence of ordinal comparison for dependent discrete event dynamic systems. Journal of Optimization Theory and Applications 94(1):29–54

David HA, Nagaraja HN (2003) Order statistics, 3rd edn, Wiley Series in Probability and Statistics. Wiley-Interscience, New York, NY

De Jong KA (1975) An analysis of the behavior of a class of genetic adaptive systems. Ph.D. thesis, University of Michigan

Deng M, Ho YC (1997) Iterative ordinal optimization and its applications. In: Proceedings of the 36th IEEE Conference on Decision and Control, vol 4, pp 3562– 3567

Deng M, Ho YC (1999) Sampling-selection method for stochastic optimization problems. Automatica 35(2):331–338

Deng M, Ho YC, Hu JQ (1992) Effect of correlated estimation errors in ordinal optimization. In: Swain JJ, Goldsman D, Crain RC, Wilson JR (eds) Proceedings of the 1992 Winter Simulation Conference pp 466–474

Dorigo M (1992) Optimization, learning and natural algorithms. Ph.D. thesis, Politecnico di Milano, Italy

Dorigo M, Di Caro G (1999) The ant colony optimization meta-heuristic. In: Corne D, Dorigo M, Glover F (eds) New ideas in optimization. McGraw-Hill, London, UK, pp 11–32

Dorigo M, Stützle T (2004) Ant colony optimization. MIT Press, Cambridge, MA

Dorigo M, Maniezzo V, Colorni A (1996) The ant system: Optimization by a colony of cooperating agents. IEEE Transactions on Systems, Man, and Cybernetics – Part B 26(1):29–41

Dorigo M, Di Caro G, Gambardella LM (1999) Ant algorithms for discrete optimization. Artificial Life 5(2):137–172

Driankov D, Hellendoorn H, Reinfrank M (2006) An introduction to fuzzy control, 2nd edn. Springer, New York, NY

Eglese RW (1990) Simulated annealing: A tool for operational research. European Journal on Operational Research 46:271–281

Falcioni M, Deem W (2000) Library design in combinatorial chemistry by Monte Carlo methods. Physical Review E 61(5):5948–5952

Falkenauer E, Bouffoix S (1991) A genetic algorithm for job-shop. In: Proceedings of the 1991 IEEE International Conference on Robotics and Automation, pp 824–829

Fishman GS (1996) Monte Carlo: Concepts, algorithms, and applications. Springer, Berlin Heidelberg New York

Floudas CA, Pardalos PM, Adjiman CS, Esposito WR, Gumus Z, Harding ST, Klepeis JL, Meyer CA, Schweiger CA (1999) Handbook of test problems for local and global optimization. Kluwer Academic Publishers, the Netherlands

Fu MC, Jin X (2001) On the convergence rate of ordinal comparisons of random variables. IEEE Transactions on Automatic Control 46(12):1950–1954

Fujimoto RM (1990) Parallel discrete event simulation. Communication of The ACM 33(10):31–53

Ganz A, Wang XD (1994) Efficient algorithm for virtual topology design in multihop lightwave networks. IEEE/ACM Transactions on Networking 2(3): 217–225

Gelenbe E, Mitrani I (1980) Analysis and synthesis of computer systems. New York Academic Press

Gentle JE (2003) Random number generation and Monte Carlo methods, 2nd edn. Springer, New York, NY

Glasserman P, Vakili P (1992) Correlation of uniformized Markov chains simulated in parallel. In: Proceedings of the Winter Simulation Conference, pp. 412–419

Glasserman P, Yao D (1992) Some guidelines and guarantees for common random numbers. Management Science 38(6):884–908

Glover F (1986) Future paths for integer programming and links to artificial intelligence. Computers and Operations Research 13:533–549

Glover F (1989) Tabu search, Part I. ORSA Journal on Computing 1(3):190–206

Glover F (1990) Tabu search, Part II. ORSA Journal on Computing 2:4–32

Glover F, Laguna M (1997) Tabu search. Kluwer Academic Publishers, Norwell, MA

Goldberg DE (1989) Genetic algorithms in search, optimization, and machine learning. Addison-Wesley, Reading, MA

Goldberg DE, Lingle R (1985) Alleles, loci and traveling salesman problem. In: Proceedings of the International Conference on Genetic Algorithms and their Applications, Carnegie-Mellon University

Goldsman D, Nelson B (1994) Ranking, selection, and multiple comparison in computer simulation. In: Proceedings of the 1994 Winter Simulation Conference

Guan X, Ho YC, Lai F (2001) An ordinal optimization based bidding strategy for electric power suppliers in the daily energy market. IEEE Transactions on Power Systems 16(4):788–797

Guide VDR Jr (2000) Technical note: Production planning and control for remanufacturing: Industry practice and research needs. Journal of Operations Management 18:467–483

Guide VDR Jr, Jayaraman V, Srivastava R (1999) Production planning and control for remanufacturing: A state-of-the-art survey. Robotics and Computer Integrated Manufacturing 15:221–230

Gunasekera JS, Fischer CE, Malas JC, Mullins WM, Yang MS, Glassman N (1996) The development of process models for use with global optimization of a manufacturing system. ASME 1996 International Mech. Eng. Conference and Exposition, Atlanta, GA

Gupta SS, Panchapakesan S (1979) Multiple decision procedures: Theory and methodology of selecting and ranking populations. Wiley, New York, NY

Harvard Center for Textile and Apparel Research Annual Progress Report, July 1 1994 to Oct 30 1995

Heidelberger P, Meketon M (1980) Bias reduction in regenerative simulation. IBM, New York, Research Report RC 8397

Ho YC (1989) Introduction to special issue on dynamics of discrete event systems: In: Proceedings of the IEEE, vol 77, issue 1, pp 3–6

Ho YC (ed) (1991) Discrete event dynamic systems. IEEE Press

Ho YC (1999) An explanation of ordinal optimization: Soft computing for hard problems. Information Sciences 113(3-4):169–192

Ho YC (2005) On centralized optimal control. IEEE Transactions on Automatic Control 50(4):537–538

Ho YC, DEDS Group (1992) Parallel discrete event dynamic system simulation. Technical Report, DEDS Group, MasPar Challenge Report

Ho YC, Pepyne D (2004) Conceptual framework for optimization and distributed intelligence. In: Proceedings of the 43rd IEEE Conference on Decision and Control, pp 4732–4739

Ho YC, Sreenivas R, Vakili P (1992) Ordinal optimization of discrete event dynamic systems. Journal of Discrete Event Dynamic Systems 2(2):61–88

Ho YC, Zhao QC, Pepyne DL (2003) The no free lunch theorems: complexity and security. IEEE Transactions on Automatic Control 48(5):783–793

Holland JH (1975) Adaptation in natural and artificial systems. University of Michigan Press, Ann Arbor, MI

Hsieh BW, Chen CH, Chang SC (2001) Scheduling semiconductor wafer fabrication by using ordinal optimization-based simulation. IEEE Transactions on Robotics and Automation 17(5):599–608

Jia QS (2006) Enhanced ordinal optimization: A theoretical study and applications. Ph.D. thesis, Tsinghua University, Beijing, China

Jia QS, Ho YC, Zhao QC (2004) Comparison of selection rules for ordinal optimization. Technical report, CFINS, Tsinghua University and Harvard University, also available online: http://cfins.au.tsinghua.edu.cn/personalhg/jiaqingshan/JiaHoZhaoSR2004.pdf.

Jia QS, Ho YC, Zhao QC (2006a) Comparison of selection rules for ordinal optimization. Mathematical and Computer Modelling, Special Issues on Optimization and Control for Military Applications 43(9-10):1150–1171

Jia QS, Zhao QC, Ho YC (2006b) A method based on Kolmogorov complexity to improve the efficiency of strategy optimization with limited memory space. In: Proceedings of the 2006 American Control Conference, Minneapolis, MN, June 14-16, pp 3105–3110

Johnston DS, Aragon CR, McGeoch LA (1989) Optimization by simulated annealing: an experiment evaluation: Part 1, graph partitioning. Operational Research 37:865–892

Kirkpatrick S, Gelatt CD Jr, Vecci MP (1983) Optimization by simulated annealing. Science 22:41–56

Kleijn MJ, Dekker R (1998) An overview of inventory systems with several demand classes. Econometric Institute Report 9838/A

Kleinrock L (1975) Queueing Systems, vol I: Theory. John Wiley and Sons, New York, NY

Kokotovic P (1992) The joy of feedback: nonlinear and adaptive. IEEE Control System Magazine 12(3):7–17

Kolmogorov A (1965) Three approaches to the quantitative definition of information. Problems of Information Transmission 1(1):1–7

Koulamas C, Anotony SR, Jean R (1994) A survey of simulated annealing application to operations research problems. International Journal of Production Research 30(1):95–108

Ku HM, Karimi L (1991) An evaluation of simulated annealing for batch process scheduling. Industrial and Engineering Chemistry Research 30:163–169

Kuehn PJ (1979) Approximation analysis of general queueing networks by decomposition. IEEE Transactions on Communications 27(1):113–126

L'Ecuyer P (2004) Random number generation. In: Chapter 2 of Gentle JE, Haerdle W, Mori Y (eds) the Handbook of Computational Statistics. Springer-Verlag, New York, NY, pp 35–70

Landau DP, Binder K (2000) A guide to Monte Carlo simulations in statistical physics. Cambridge University Press, Cambridge

Lau TWE, Ho YC (1997) Alignment probabilities and subset selection in ordinal optimization. Journal of Optimization and Application 93(3):455–489

Law AM, Kelton WD (1991) Simulation modeling and analysis, McGraw-Hill Inc., New York, NY

Lee LH (1997) Ordinal optimization and its application in apparel manufacturing systems. Ph.D. thesis, Harvard University

Lee LH, Lau TWE, Ho YC (1999) Explanation of goal softening in ordinal optimization. IEEE Transactions on Automatic Control 44(1):94–99

Lee JT, Lau EL, Ho YC (2001) The Witsenhausen counterexample: A hierarchical search approach for nonconvex optimization problems. IEEE Transactions on Automatic Control 46(3):382–397

Levin LA (1973) Universal sequential search problems. Problems of Information Transmission 9(3):265–266

Levin LA (1984) Randomness conservation inequalities: Information and independence in mathematical theories. Information and Control 61:15–37

Li M, Vitányi P (1997) An introduction to Kolmogorov complexity and its applications, 2nd edn. Springer, Berlin Heidelberg New York

Lin X (2000a) A discussion on performance value versus performance order. IEEE Transactions on Automatic Control 45(12):2355–2358

Lin X (2000b) A new approach to discrete stochastic optimization problems. Ph.D. thesis, Harvard University

Lin J (2004) A research on simulation budget allocation and its application for optimizing the reliability of transportation system capacity. Ph.D. thesis, University of Pennsylvania

Lin SY, Ho YC (2002) Universal alignment probability revisited. Journal of Optimization Theory and Applications 113(2):399–407

Lin SY, Ho YC, Lin CH (2004) An ordinal optimization theory-based algorithm for solving the optimal power flow problem with discrete control variables. IEEE Transactions on Power Systems 19(1):276–286

Luh PB, Liu F, Moser B (1999) Scheduling of design projects with uncertain number of iterations. European Journal of Operational Research 113:575–592

Luo YC (2000) Very efficient hybrid simulation methods for complex problems. Ph.D. thesis, University of Pennsylvania

Luo YC, Guignard M, Chen CH(2001) A hybrid approach for integer programming combining genetic algorithms, linear programming and ordinal optimization. Journal of Intelligent Manufacturing 12:509–519

Metropolis N, Rosenbluth A, Teller A, Teller E (1953) Equation of state calculations by fast computing machines. Journal of Chemical Physics 21:1087–1092

Mori H, Tani H (2003) A hybrid method of PTS and ordinal optimization for distribution system service restoration. In: Proceedings of the 2003 IEEE International Conference on Systems, Man and Cybernetics, Washington, DC, pp 3476–3483

Nawaz M, Enscore E, Ham I (1983) A heuristic algorithm for the m-machine, n-job flow-shop sequencing problem. Omega 11(1):91–95

Nelson BL, Swann J, Goldsman D, Song W (2001) Simple procedures for selecting the best simulated system when the number of alternatives is large. Operations Research 49:950–963

Ólafsson S (1999) Iterative ranking-and-selection for large-scale optimization. In: Proceedings of the 1999 Winter Simulation Conference, pp 479–485

Ólafsson S, Shi L (1999) Optimization via adaptive sampling and regenerative simulation. In: Farrington PA, Nembhard HB, Sturrock DT, Evans GW (eds) Proceedings of the 1999 Winter Simulation Conference, pp 666–672 OR-Library, http://www.ms.ic.ac.uk/info.html

Painton LA, Diwekar UM (1994) Synthesizing optimal design configurations for a brayton cycle power plant. Computers and Chemical Engineering 18(5):369–381

Papadimitriou CH, Tsitsiklis JN (1986) Intractable problems in control theory. SIAM Journal on Control and Optimization 24(4):639–654

Pareto V (1896) Cours d'Economie Politique. Rouge, Lausanne, Switzerland

Passino KM, Yurkovich S (1998) Fuzzy Control. Addison Wesley Longman, Menlo Park, CA

Patsis NT, Chen CH, Larson ME (1997) SIMD parallel discrete-event dynamic system simulation. IEEE Transactions on Control Systems Technology 5(1):30–41

Reeves CR (1995) A genetic algorithm for flowshop sequencing. Computers and Operations Research 22(1):5–13

Ren Z, Krogh BH (2002) State aggregation in Markov decision processes. In: Proceedings of the 41st IEEE Conference on Decision and Control, Las Vegas, NV, December 10-13, pp 3819–3824

Rose C, Smith MD (2002) Order Statistics. In: Section 9.4 of Mathematical statistics with Mathematica. Springer-Verlag, New York, NY, pp 311–322

Santer TJ, Tamhane AC (1984) Design of experiments: Ranking and selection. Marcel Dekker, New York, NY

Schainker R, Miller P, Dubbelday W, Hirsch P, Zhang G (2006) Real-time dynamic security assessment: fast simulation and modeling applied to emergency outage security of the electric grid. IEEE Power and Energy Magazine 4(2):51–58.

Schwartz B (2004) The paradox of choice – why more is less, Harper Collins, New York, NY

Shedler GS (1993) Regenerative stochastic simulation. Academic Press Inc., Burlington, MA

Shen Z, Bai HX, Zhao YJ (2007) Ordinal optimization references list, May 1, http://www.cfins.au.tsinghua.edu.cn/uploads/Resources/Complete_Ordinal _Optimization_Reference_List_v7.doc

Shi L, Chen CH (2000) A new algorithm for stochastic discrete resource allocation optimization. Discrete Event Dynamic Systems: Theory and Applications 10:271–294

Shi L, Chen CH, Yücesan E (1999) Simultaneous simulation experiments and nested partitions for discrete resource allocation in supply chain management. In: Proceedings of the 1999 Winter Simulation Conference, pp 395–401

Shi L, Ólafsson S (1998) Hybrid equipartitioning job scheduling policies for parallel computer systems. In: Proceedings of the 37th IEEE Conference on Decision and Control, pp 1704–1709

Shi L, Ólafsson S (2000a) Nested partitions method for global optimization. Operations Research 48:390–407

Shi L, Ólafsson S (2000b) Nested partitions method for stochastic optimization. Methodology and Computing in Applied Probability 2:271–291

Song C, Guan X, Zhao Q, Ho YC (2005a) Machine learning approach for determining feasible plans of a remanufacturing system. IEEE Transactions on Automation Science and Engineering [see also IEEE Transactions on Robotics and Automation] 2(3):262–275

Song C, Guan X, Zhao Q, Jia Q (2005b) Planning remanufacturing systems by constrained ordinal optimization method with feasibility model. In: the 44th IEEE Conference on Decision and Control and European Control Conference, Seville, Spain, December 12-15, pp 4676–4681

Specht DF (1991) A general regression neural network. IEEE Transactions on Neural Network 2(6):568–576

Stadler PF, Schnabl W (1992) The landscape of the traveling salesman problem. Physics Letters A 161:337–344

Sullivan KA, Jacobson SH (2000) Ordinal hill climbing algorithms for discrete manufacturing process design optimization problems. Discrete Event Dynamic Systems: Theory and Applications 10:307–324

Tan KC, Lee TH, Khor EF (2002) Evolutionary algorithms for multiobjective optimization: Performance assessments and comparisons. Artificial Intelligence Review 17:253–290

Tezuka S (1995) Uniform random numbers: Theory and practice. Kluwer Academic Press, Boston, MA

Thomke SH (2003) Experimentation matters: Unlocking the potential of new technologies for innovation, Harvard Business School Press, Boston, MA

Tyan HY (2002) Realization and evaluation of a computer-based compositional software architecture for network simulation. Ph.D. thesis, The Ohio State University

Vakili P (1991) A standard clock technique for efficient simulation. Operations Research Letters 10:445–452

Vakili P, Mollamustafaoglu L, Ho YC (1992) Massively parallel simulation of a class of discrete event systems. In: Proceedings of the IEEE Symposium on the Frontier of Massively Parallel Computation

Van Laarhoven PJM, Aarts EHL (1987) Simulated annealing: Theory and applications. D. Reidel Publishing, Dordrecht, The Netherlands

Vertosick F (2002) The genius within: Discovering the intelligence of every living thing, Harcourt

Volpe AP (2005) Modeling flexible supply options for risk-adjusted performance evaluation. Ph.D. thesis, Harvard University

Walker AJ (1974) New fast method for generating discrete random numbers with arbitrary distributions. Electronic Letters 10(8):127–128

Whitt W (1980) Continuity of Generalized Semi-Markov Processes. Mathematics of Operations Research 5(4):494–501

Whitt W (1983) The queueing network analyzer. The Bell System Technical Journal 62(9):2779–2815

Wieselthier JE, Barnhart CM, Ephremides A (1995) Standard clock simulation and ordinal optimization applied to admission control in integrated communication-networks. Discrete Event Dynamic Systems: Theory and Applications 5(2-3):243–280

Wilson JR (2001) A multiplicative decomposition property of the screening-and-selection procedures of Nelson et al. Operations Research 49(6):964–966

Witsenhausen HS (1968) A counterexample in stochastic optimum control. SIAM Journal on Control 6(1):131–147

Xia L, Zhao Q, Jia QS (2004) The SRLF rule in multi-unit joint replacement maintenance problem and its optimality. The 12th INFORMS/APS Conference, Beijing, China, June 23-25

Xie X (1994) Ordinal optimization approach to a token partition problem for stochastic timed event graphs. In: Tew JD, Manivannan S, Sadowski DA, Seila AF (eds) Proceedings of the 1994 Winter Simulation Conference, Orlando, FL, December 11-14, pp 581–588

Xie XL (1997) Dynamics and convergence rate of ordinal comparison of stochastic discrete-event systems. IEEE Transactions on Automatic Control 42(4):586–590

Yakowitz SJ (1977) Computational probability and simulation. Addison-Wesley, Advanced Book Program, Reading, MA

Yang MSY (1998) Ordinal optimization and its application to complex deterministic problems. Ph.D. thesis, Harvard University

Yang MS, Lee LH (2002) Ordinal optimization with subset selection rule. Journal of Optimization Theory and Applications 113(3):597–620

Yang MS, Lee LH, Ho YC (1997) On stochastic optimization and its applications to manufacturing. In: Proceedings of the AMS-SIAM Summer Seminar The Mathematics of Stochastic Manufacturing Systems, Williamsburg, VA, pp 317–331

Yen CH, Wong DSH, Jang SS (2004) Solution of trim-loss problem by an integrated simulated annealing and ordinal optimization approach. Journal of Intelligent Manufacturing 15:701–709

Yoo T, Kim D, Cho H (2004) A new approach to multi-pass scheduling in shop floor control. In: Proceedings of the 2004 Winter Simulation Conference, pp 1109–1114

Zhang L (2004) Study on order-based intelligent algorithms for simulation optimization problems. Master thesis, Tsinghua University, Beijing, China

Zitzler E, Thiele L, Laumanns M, Fonseca CM, da Fonseca VG (2003) Performance assessment of multiobjective optimizers: An analysis and review. IEEE Transactions on Evolutionary Computation 7:117–132

Index